Changing Climate, Changing Economy

Cournot Centre for Economic Studies

Series Editor: Robert M. Solow, *Emeritus Professor of Economics, Massachusetts Institute of Technology; President of the Cournot Centre for Economic Studies, Paris, France*

Conference Editor: Jean-Philippe Touffut, *Director of the Cournot Centre for Economic Studies, Paris, France*

Changing Climate, Changing Economy

Edited by

Jean-Philippe Touffut

Director of the Cournot Centre for Economic Studies, Paris, France

THE COURNOT CENTRE FOR ECONOMIC STUDIES SERIES

Edward Elgar

Cheltenham, UK • Northampton, MA, USA

Published by
Edward Elgar Publishing Limited
The Lypiatts
15 Lansdown Road
Cheltenham
Glos GL50 2JA
UK

Edward Elgar Publishing, Inc.
William Pratt House
9 Dewey Court
Northampton
Massachusetts 01060
USA

A catalogue record for this book
is available from the British Library

Library of Congress Control Number: 2009933413

Mixed Sources
Product group from well-managed
forests and other controlled sources
www.fsc.org Cert no. SA-COC-1565
© 1996 Forest Stewardship Council

ISBN 978 1 84844 836 0 (cased)
ISBN 978 1 84844 837 7 (paperback)

Printed and bound by MPG Books Group, UK

Contents

Figures and boxes

FIGURES

BOX

Contributors

Michel Armatte is a Researcher at the Centre Alexandre Koyré – Centre de Recherche en Histoire des Sciences et des Techniques du CNRS (Centre National de la Recherche Scientifique) in Paris. He is an Associate Professor and Director of the Department of Applied Economics at the University Paris-Dauphine. His research interests include mathematical modelling in social sciences and the history of statistics and econometrics.

Jean-Pierre Dupuy is a Professor of Social and Political Philosophy at the École Polytechnique, a Research Director in Philosophy at the CNRS (Centre National de la Recherche Scientifique), and a Professor of French and Political Science at Stanford University, where he is a Researcher at the Center for the Study of Language and Information. His current research focuses on the paradoxes of rationality, analytic philosophy, and cognitive science; the ethics of nuclear deterrence and preemptive war; and the philosophy of risk and uncertainty.

Olivier Godard is a Senior Researcher at the CNRS (Centre National de la Recherche Scientifique) and a Professor of the Socio-economics of Sustainable Development at the École Polytechnique. He was a regular consultant for the OECD Environment Directory on the use of economic instruments for environmental policies in the 1990s. His main fields of research include the philosophy of economics, environmental economics, and policies under uncertainty and controversy.

Inge Kaul is a Professor at the Hertie School of Governance in Berlin and acts as a policy adviser to various think-tanks and international organizations. She was the first Director of the United Nations Development Programme (UNDP) Human Development Report Office, a position that she held from 1989 to 1994. She was Director of the UNDP's Office of Development Studies until 2005.

Thomas Schelling is Distinguished University Professor Emeritus at the University of Maryland and Professor Emeritus of Political Economy at Harvard University. As a former President of the American Economic Association, he is also a member of the US National Academy of Sciences and the American Academy of Arts and Sciences. His diverse research interests include energy and environmental policy, climate change, as well as conflict and bargaining theory, for which he was awarded the Nobel Memorial Prize in Economics in 2005.

Robert M. Solow is Institute Professor Emeritus at the Massachusetts Institute of Technology. In 1987, he was awarded the Nobel Memorial Prize in Economics for his contributions to economic growth theory. He is a Visiting Scholar at the Russell Sage Foundation, New York, where he is a member of the advisory committee for the Foundation's project on the incidence and quality of low-wage employment in Europe and the United States. Professor Solow is President of the Cournot Centre for Economic Studies.

Nicholas Stern is a Professor of Economics and Government and Chair of the Grantham Research Institute on Climate Change and the Environment at the London School of Economics. He was Head of the Stern Review on the Economics of Climate Change and the British Government Economic Service. He was chief economist at both the World Bank and the European Bank for Reconstruction and Development. His research focuses on economic development and growth, economic theory, tax reform, public policy and the role of the state and economies in transition.

Thomas Sterner is a Professor of Environmental Economics at the University of Gothenburg and a University Fellow at Resources for the Future (RFF). He founded the unit for environmental economics in Gothenburg and has participated in its development since 1990. He is currently President of the European Association of Environmental and Resource Economists. His main research interests lie in the design of policy instruments as applied to energy and climate, natural resource management, and industrial and transport pollution.

Jean-Philippe Touffut is Co-Founder and Director of the Cournot Centre for Economic Studies.

Martin L. Weitzman is a Professor of Economics at Harvard University. He is a Fellow of the Econometric Society and the American Academy of Arts and Sciences. His current research focuses on environmental economics, including climate change, the economics of catastrophes, cost–benefit analysis, long-run discounting, green accounting, and comparison of alternative instruments for controlling pollution.

Preface

This volume is one of a series arising from the conferences organized by the Cournot Centre for Economic Studies, Paris. These conferences explore contemporary issues in economics, with particular focus on Europe. Speakers, along with other participants and members of the audience, are drawn from backgrounds in academia, business, finance, labour unions, the media and national or multinational governmental and non-governmental agencies.

The contributions of this book originated from the eleventh conference of the Cournot Centre for Economic Studies held on 18 and 19 December 2008.

Acknowledgements

I would like to thank all the participants of the Cournot Centre conference on the economics of climate change for their contribution to this volume, and, in particular, Robert Solow for his continued support and guidance of the Centre and its activities.

My heartfelt thanks go to Therrese Goodlett and Anna Kaiser. From the organization of the conference to the preparation of the manuscript, they have enabled this work to see the light of day under the best possible conditions. A very special thanks also goes to Richard Crabtree for translations and transcription work. A film of the whole conference is available on the Cournot Centre's website: www.centrecournot.org.

About the series

Professor Robert M. Solow

The Cournot Centre for Economic Studies is an independent, French-based research institute. It takes its name from the pioneering economist, mathematician and philosopher Antoine Augustin Cournot (1801–77).

Neither a think-tank nor a research bureau, the Centre enjoys the special independence of a catalyst. My old student dictionary (dated 1936) says that catalysis is the 'acceleration of a reaction produced by a substance, called the *catalyst*, which may be recovered practically unchanged at the end of the reaction'. The reaction we have in mind results from bringing together (a) an issue of economic policy that is currently being discussed and debated in Europe and (b) the relevant theoretical and empirical findings of serious economic research in universities, think-tanks and research bureaux. Acceleration is desirable because it is better that reaction occurs before minds are made up and decisions taken, not after. We hope that the Cournot Centre can be recovered practically unchanged and used again and again.

Notice that 'policy debate' is not exactly what we are trying to promote. To have a policy debate, you need not only knowledge and understanding, but also preferences, desires, values and goals. The trouble is that, in practice, the debaters often have only those things, and they invent or adopt only those 'findings' that are convenient. The Cournot Centre hopes to inject the findings of serious research at an early stage.

It is important to realize that this is not easy or straightforward. The analytical issues that underlie economic policy choices are usually complex. Economics is not an experimental science. The available data are scarce, and may not be exactly the relevant ones. Interpretations are therefore uncertain. Different studies, by uncommitted economists, may give different results. When those controversies exist, it is our hope that the Centre's conferences will discuss them. Live debate at that fundamental level is exactly what we are after.

There is also a problem of timing. Conferences have to be planned well in advance, so that authors can prepare careful and up-to-date texts. Then a publication lag is inevitable. The implication is that the Cournot Centre's conferences cannot take up very short-term issues of policy. Instead, a balancing act is required: we need issues that are short-term enough so that they are directly concerned with current policy, but long-term enough so that they remain directly relevant for a few years.

I used the words 'serious research' a moment ago. That sort of phrase is sometimes used to exclude unwelcome ideas, especially unfashionable ones. The Cournot Centre does not intend to impose narrow requirements of orthodoxy, but it does hope to impose high standards of attention to logic and respect for facts. It is because those standards are not always observed in debates about policy that an institution like the Cournot Centre has a role to play.

OTHER BOOKS IN THE COURNOT CENTRE SERIES

2009. *Does Company Ownership Matter?*
Edited by Jean-Philippe Touffut
Contributors: Jean-Louis Beffa, Margaret Blair, Wendy Carlin, Christophe Clerc, Simon Deakin, Jean-Paul Fitoussi, Donatella Gatti, Gregory Jackson, Xavier Ragot, Antoine Rebérioux, Lorenzo Sacconi, Robert M. Solow

2008. *Central Banks as Economic Institutions*
Edited by Jean-Philippe Touffut
Contributors: Patrick Artus, Alan Blinder, Willem Buiter, Barry Eichengreen, Benjamin Friedman, Carl-Ludwig Holtfrerich, Gerhard Illing, Otmar Issing, Takatoshi Ito, Stephen Morris, André Orléan, Nouriel Roubini, Robert M. Solow

2007. *Augustin Cournot: Modelling Economics*
Edited by Jean-Philippe Touffut
Contributors: Robert J. Aumann, Alain Desrosières, Jean Magnan de Bornier, Thierry Martin, Glenn Shafer, Robert M. Solow, Bernard Walliser

2006. *Advancing Public Goods*
Edited by Jean-Philippe Touffut
Contributors: Patrick Artus, Avner Ben-Ner, Bernard Gazier, Xavier Greffe, Claude Henry, Philippe Herzog, Inge Kaul, Joseph E. Stiglitz

2005. *Corporate Governance Adrift: A Critique of Shareholder Value*
Michel Aglietta and Antoine Rebérioux

2004. *The Future of Economic Growth: As New Becomes Old*
Robert Boyer

2003. *Institutions, Innovation and Growth: Selected Economic Papers*
Edited by Jean-Philippe Touffut
Contributors: Philippe Aghion, Bruno Amable, with Pascal Petit, Timothy Bresnahan, Paul A. David, David Marsden, AnnaLee Saxenian, Günther Schmid, Robert M. Solow, Wolfgang Streeck, Jean-Philippe Touffut

Introduction: changing climate, changing economists?

Jean-Philippe Touffut

Climate issues have only recently become of interest to disciplines outside climatology. Up until the late 1970s, historians, geographers and economists rarely touched on the subject.[1] In 1988, politicians seized the occasion of a meeting of the G7 and the General Assembly of the United Nations to introduce it into their discourse and to officially solicit economic expertise on the matter. Ten years later, 172 countries signed the Kyoto Protocol, setting a timetable for reducing the emission of six greenhouse gases, together considered to be the main cause of global warming over the last 50 years. The role of economists has grown constantly since then, to such a point that their discourse seems to have gained equal pertinence to that of climatologists on the subject today (Hirshleifer, 1985). Economists do not generally work in a hybrid manner, as their approach to the fields of sociology, anthropology and law bears witness. They typically adopt their economic methods and apply them to their subjects of study. This volume is not only another occasion for climate economists to develop their analyses and formulate recommendations; it is also an opportunity to reconsider all that is at stake with their appropriation of this field.

Ten years after the signing of the Protocol, the results obtained remain a far cry from the exhortations of Kyoto. What was the role of economics in the scientific and political debates of that period? Although at the origin of international negotiations leading up to the Protocol, climatologists quickly found themselves pitted against economists whose authority in the field no longer seemed to be questioned. Economists' positions on climate change have become dominant. In general, they are hostile to the rapid reduction of emissions; they usually favour applying taxes. How has economic discourse come to have such legitimacy in the climate debates?

A first response can be found in the probabilistic approach and statistical methods that economists diffuse at the political level, and which appear to be so user friendly to politicians. Serving as a governmental tool since the eighteenth century, statistics has become the 'instrument of proof' since the nineteenth century and its closer linkage with probability theory

1

(Foucault, 2004; Desrosières, 2008). In climate analysis, statistics is used only after the climate models have produced enough results and the data have been registered in a workable form.

In the 1950s and 1960s, weather forecasting was explored using physics-based models describing the movement of the atmosphere and the water cycle from ocean-surface evaporation to the formation of precipitation. Developed with a view to meteorological applications, these models were then applied to climatic time scales of a month and more. The forecasts involved average values rather than the monitoring of individual per-turbations. The first experiments conducted in the 1960s confirmed that these models were capable of simulating the main characteristics of the climate. Whereas weather is measured in terms of temperature, pressure, precipitation, wind, dew-point or visibility, climate is described in terms of averages, frequencies and extremes. Climate data are therefore avail-able in the form of statistical series. Economists take part in the process of quantification of weather phenomena by expressing in numerical form what had been expressed in words. This is not a neutral transformation. It requires the prior creation and explicit formulation of equivalence con-ventions, involving comparisons, codified and reproducible negotiations and calculations leading to the numerical expression. Measurement in the strict sense of the term comes afterwards, as the regulated implementation of those conventions. Things are supposed to exist in measurable form, like the height of a mountain. Measurement is the second stage of quan-tification, once the convention has been established among researchers (Desrosières, 2006).

Economic discourse is influential, because it is predisposed to norma-tivity and adapts easily to one of the key factors of climate study: uncer-tainty, a concept about which economists have constructed a powerful body of literature (Pradier, 2006). In the field of climate study, uncertainty has certainly not diminished: when asked how the surface temperature of the planet is going to evolve, environmental scientists give more or less the same answer today as the US National Academy of Sciences did in the 1970s. A doubling in the concentration of CO_2 could lead to a rise in tem-perature of somewhere between 1.5 and 4.5°C. In one sense, as Thomas Schelling points out in Chapter 1, uncertainty is in fact greater now, for two reasons: first, because it takes a good deal of courage to propose esti-mations that diverge from the ones that have been accepted for the last 30 years, and second, because the mechanisms of climate have turned out to be much more complicated than expected. The factor of 3 (from 1.5 to 4.5°C), by which estimations vary, stems largely from the way the role of the oceans and clouds is dealt with. Thomas Sterner (Chapter 4) reminds us that not everything is uncertain, such as the impact of greenhouse gases on temperature increases. The highest degree of uncertainty, however, is found in the economic costs of climate change.

To what extent do the concepts and methods developed in economics offer the appropriate analytic tools for understanding the situation and the best normative benchmarks for determining action? A detour through the history of economic thought can help answer that question. Olivier Godard (Chapter 2) invites us to look at economists' ambition to discipline the climate study fields using the most general concepts and models, which were created in other contexts to resolve problems such as price information, market imperfections or asset pricing. From a mainstream view, the environmental issue can be considered as nothing more than a problem of restoring optimality in the allocation of goods between agents with incompatible demands. There exist at least two branches of economics that do not agree with this standard analysis: ecological economics and socio-economics. Nonetheless, it is the recommendations of the dominant economic worldview that are heard over these branches and adopted by policy makers.

The contrast is strong between the time scale – a half-century or more – of such policy recommendations and growth. This challenge for researchers, used to basing their studies on a production–consumption cycle of goods and services in a market in which supply and demand adjust quickly, forces economists to find an appropriate framework in which they can build and develop their approach. Michel Armatte (Chapter 3) questions the methodology of scenarios. Overall, they answer the needs of the scientific, political and NGO communities in terms of both the production of assessments and the negotiation of mitigation and adaptation policies. The division of labour between these communities responds to a demand for policy assessment, which prevails over purely epistemological rationales and over the organization of the supply of scenarios, models and results to meet that demand.

Scenarios decontextualize the climate question, reducing it to issues of calculation to such an extent that the policy recommendations it produces are in some way 'locked' in to the economic argument. Economists, for instance, discount future costs and benefits at a positive rate. Reducing global carbon emissions or investing in technologies for mitigating global warming would involve huge costs now, but the benefits from averting economic disruptions would only be enjoyed in 50 years' time or more. When economists evaluate public projects, they typically use long-term interest rates on government bonds to discount future benefits and costs, which they consider to be the 'opportunity cost of capital'. The expression refers to the rate of interest that could be earned by investing in government bonds rather than in the project whose benefits and costs are being evaluated. If you discount, for instance, at 4 per cent a year, a dollar's worth of additional consumption benefits 100 years from now would be worth less than 3 cents today. This is another way of saying that as the price for giving up one dollar's worth of consumption today, you would expect to have more than 30 dollars' worth of consumption benefits available 100 years from now. Economic models of climate change have shown that if you use an annual discount rate of, say, 4 per cent, the negative benefits are greater than the sum of the discounted benefits from curbing net carbon emissions. Doing something about climate change now, the calculations imply, would be to throw money away.

Discounting appears to suppress the concerns raised by environmentalists involving future environmental damages caused by human activities today. This has led some people to reject the concept of discounting, or more generally of cost–benefit analysis. Thomas Sterner challenges that view in Chapter 4 by demonstrating that the problem is not discounting itself, but how it is used. He brings to the fore that discounting contains two fundamental questions involving economic analysis: will growth – hidden behind this coefficient – be possible for centuries to come? How will prices behave? Generally speaking, the discounting rate mirrors the image we have of our future, and notably of our confidence level, or catastrophic posture. On this subject, Martin Weitzman (Chapter 5) places us precisely in a situation where relative risk aversion is positive (a hypothesis shared by all the integrated assessment models), and where climate change could generate a catastrophe, without us knowing the objective probabilities of that catastrophe occurring. The latter is verified, because there is no history (on the scale of several million years) of the CO_2 concentrations reaching the levels that they have today. Weitzman shows that all the conclusions of cost–benefit analyses – that we should wait before acting to strongly reduce emissions – are invalidated. This result is of capital importance.

How can we determine the level of concentration of greenhouse gases below which this conclusion – the 'dismal theorem' – no longer applies? Several authors have sought to define the optimal form that climate policy should take, particularly as regards the choice between taxes and quotas. These papers, based on a previous article by Weitzman (1974), conclude that taxes are better than quotas, because they would allow us to adapt the level of emissions if predictions of the cost of emissions reduction turn out to be inaccurate. In particular, taxes are still preferable even in the presence of a catastrophe if we do not know at what level of CO_2 concentrations the catastrophe might occur (Pizer, 2003). These works are nevertheless based on probability density functions without the 'fat tails', which are, in the above-cited article by Weitzman, due to the absence of objective probabilities. Would taking into account this factor alter the conclusion that taxes are better than quotas?

Economics does more than just formulate policy recommendations, it emphasizes who the main victims of global warming will be: those concentrated in the developing world. As a result, it is difficult to mobilize the main actors responsible for climate change, because they do not feel threatened over the short term by the consequences. The crucial issue then, as Schelling points out, is how to make the recommendations agreed by the international community enforceable. On this point, it may be that we are currently witnessing a turning point in the attitude of economists. As Weitzman says, the important thing is that the authors reach the right conclusions, even if it is for the wrong reasons (Weitzman, 2009).

The contributions in this volume tend to converge on how economics should analyse climate change. Such unanimity is harder to find in the world at large, or in the part of the research world that is interested in this question. Differences of opinion often arise, not out of differences in the fundamental science, but out of differences in the economic analysis. In the final contribution of this book, the participants of the round table (Chapter 6) – including Inge Kaul, Thomas Schelling, Robert Solow, Nicholas Stern, Thomas Sterner and Martin Weitzman – try to draw lessons from these differences and suggest directions for future research and policy. As Nicholas Stern puts it, economists have no monopoly on the two key questions of the debate: how big are the damages of climate change, and how much do we care about the future? Inge Kaul suggests that global cost–benefit, along with disaggregated, analyses of climate change could answer these questions. It would reveal who the winners of a future global deal would be and the scope for making compensatory transfers to the potential losers. Economists and policy makers could then construct win–win bargains. Finding win–win solutions requires political will and the recognition on the part of the winners that under conditions of policy interdependence, national interests may often best be served through successful cooperative policy approaches at the international level. This could, in turn, put a stop to the hesitation – based on economic and political global inequity – that we see in coming to action.

NOTE

1. Among the first publications by economists on the subject is the 600-page report co-authored by Thomas Schelling in 1979: *Energy: The Next Twenty Years*. The report dedicated only a few pages to greenhouse gases, which have become one of the central themes in climatology today.

REFERENCES

Desrosières, A. (2006), 'From Cournot to public policy evaluation: paradoxes and controversies involving quantification', *Prisme*, no. 6, April, Cournot Centre for Economic Studies, Paris.

Desrosières, A. (2008), *Pour une sociologie historique de la quantification*, Paris: Presses de l'école des Mines.

Foucault, M. (2004), *Sécurité, territoire, population*, Cours au Collège de France, 1977–78, Paris: Hautes études/Gallimard.

Hirshleifer, J. (1985), 'The expanding domain of economics', *American Economic Review*, **75**: 53–68.

Pizer, W.A. (2003), *Climate Change Catastrophes*, Washington, DC: Resources for the Future, May, RFF DP 03-31.

Pradier, P.-C. (2006), *La Notion de risque en économie*, Paris: La découverte.

Schelling, T. and Ford Foundation Study Group (1979), *Energy: The Next Twenty Years*, edited by H. Landsberg, Cambridge, MA: Ballinger.

Weitzman, M. (1974), 'Prices vs. quantities', *Review of Economic Studies*, **41** (4), October: 477–91.

Weitzman, M. (2009), 'On modeling and interpreting the economics of catastrophic climate change', *Review of Economics and Statistics*, February, **91** (1): 1–19.

1. Climate change: a bundle of uncertainties

Thomas Schelling

This chapter is concerned with uncertainty, but let me begin with some of the certainties. If this were an American audience, I would have to explain that the concept of greenhouse gases may be, as my President used to say, 'merely a theory', but then so is gravity merely a theory. Nobody understands gravity, but it is pretty well established that things are attracted to Earth if you let go of them.

If you shine some infrared light through a glass chamber full of carbon dioxide, less light comes out than went in, because certain wavelengths in the infrared part of the spectrum are intercepted by carbon dioxide molecules. If you compare the difference between the light going in on one side and the light coming out on the other, it correlates with the rise in temperature of the gas that has intercepted the infrared radiation. That has been known for 100 years – the basic idea that so-called 'greenhouse gases' – carbon dioxide in particular – will absorb or intercept infrared radiation.

The Earth gets bathed in sunlight during the daytime, and it must get rid of that energy to avoid global warming. It gets rid of it in the form of infrared radiation. Therefore, what keeps the temperature of the atmosphere in balance is the ability of the Earth to get rid of the excess energy that comes in during the daytime. It has been understood for decades that the planet Mars, lacking any greenhouse gases, is too cold for water to exist on its surface as a liquid. It is equally known that the planet Venus is so bathed in greenhouse gases that its surface is too hot for water to exist as a liquid. Earth has a nice concentration of greenhouse gases, which allows temperatures such that water can exist as a liquid. Without liquid water, there could be no life. All life on Earth depends on liquid water, and therefore while we may lament an excess of greenhouse gases in the atmosphere, we should not forget that without carbon dioxide and associated greenhouse gases, we could not have any life on Earth.

Somebody persuaded the Supreme Court of the United States, about a year ago, to treat carbon dioxide as a pollutant. Now it has to be remembered that carbon dioxide is what all green things 'eat'. Without carbon dioxide there would be no green grass, no trees, no vegetation . . . Carbohydrates all depend on carbon dioxide, which, through photosynthesis, energized by sunlight, causes green things to grow, so that herbivores can live and carnivores can live off the herbivores, all of which depend on carbon dioxide in the atmosphere.

It is worth remembering that if we breathe carbon dioxide at many times the ordinary concentration in the atmosphere, there is no physical harm, not even to pregnant women or foetuses or small children. So it is important to bear in mind that what we call 'greenhouse gases' are absolutely essential to life on this planet, and the only problem is, if there is too much of them, we may get climate changes that we do not like. (It is worth reflecting, just for perspective, that if the concentration of oxygen in the atmosphere doubled, that would be catastrophic. Any kind of fire would become a conflagration. Both gases are essential; too much of either can become dangerous.)

A word about the term 'greenhouse'. Greenhouses do not work in the same way that the greenhouse effect works. Greenhouses merely trap warm air: we build a glass enclosure, the sunlight warms the Earth inside, which warms the air inside; the function of the greenhouse glass is simply to keep the warm air from dissipating. That is not the way the greenhouse phenomenon works.

People refer to the greenhouse problem as a 'global warming problem'; it is better to think of it as a 'climate change phenomenon'. It is not clear that every place will get warmer just because the average atmospheric surface temperature is increasing. Some places may get cooler, some places may get sunnier and some may get cloudier, some may get more precipitation and some may get less precipitation. Climate changes, and that has to be plural, not just *a* climate change, but changes in climates everywhere, are still hard to predict.

To give an example: it is hard to predict what is going to happen high up in the mountains. Not many people live above 3000 metres – Tibetans and Bolivians and a few such people – but huge numbers of people depend on what happens to the climate above 3000 metres. That is because a large part of the world's agriculture depends on moisture that falls in the form of snow in the wintertime and stays there as snow until late spring and early summer, and then melts and comes down in rivers and is available for irrigation. Most agriculture in India, Pakistan, Bangladesh, Burma, China, Chile, Peru, Argentina and California depends on precipitation in the form of snow that stays as snow until it is time to irrigate crops. If, instead of falling as snow, the precipitation falls as rain, it is largely wasted, or if it falls as snow and it melts too early, it is not available when the farmers need water. Therefore, what happens in the high mountains is important to agricultural productivity in most of the world.

Yet it is difficult for any of the current climate models to anticipate what is going to happen to climate above about 3000 metres, though it is crucially important. So instead of a global warming problem, it is a problem of climate change, of which the driving force will be an increase in the average atmospheric surface temperature that will mainly manifest itself as greater warmth.

Now, among the uncertainties, the uncertainty that gets the most attention is what is going to happen to the average atmospheric surface temperature (see Weitzman, Chapter 5). That is just a beginning uncertainty! It is interesting to look at the estimate of how much temperature change there might be with a doubling of the concentration of greenhouse gases in the atmosphere – doubling *vis-à-vis* the so-called 'pre-industrial' level of about 280 parts per million. The estimate as of the late 1970s, of the US National Academy of Sciences, was that a doubling of the concentration of CO_2 (at that time carbon dioxide was the only recognized so-called 'greenhouse gas') would lead to a change in temperature that could be anywhere between 1.5 degrees Celsius (°C) and 4.5°C. Now that is a huge range of uncertainty. It is a threefold difference: 1.5–4.5°C. If I went to my doctor and he said, 'Schelling, if you don't watch your diet, you're going to put on a lot of weight', and I said, 'How much weight am I going to put on if I don't change my diet?', and my doctor said, 'Oh, anywhere from 40 to 120 pounds', I would say 'that's quite a range of uncertainty, Doctor!'.

Now it is interesting that when the Intergovernmental Panel on Climate Change (IPCC) reviewed this estimate in 1992, it said, 'We find no reason to change that estimate', and that is still about the going estimate. Sometimes it is referred to as maybe 2–5°C.

The question is then: why has that estimate, the range of uncertainty, not been reduced in the last 30 years? The amount of money spent on research into this question, the number of people dedicating their career to studying climate change, is essentially now 1000 times what it was before the mid-1970s, and so the question arises: why has that uncertainty not been reduced?

In fact, as Martin Weitzman explains (Chapter 5), in some ways the uncertainty has increased. There are at least two reasons. One is that it takes a lot of self-confidence to offer an alternative estimate to the one that has been accepted for three decades, and nobody is quite ready to bet his/her career on a revised estimate. But more than that, there are some subjects, of which genetics and brain science are two examples, that upon examination turn out to be more complicated than people expected. Brain science now is nothing like it was thought to be 20 years ago, and genetics has changed drastically in the last couple of decades.

When I first got involved in this subject, the role of the oceans was just that of a huge cooling reservoir. We used the term 'thermal inertia', to mean that because the specific heat of water is so much greater than that of the atmosphere, you cannot get the air much warmer until you have warmed the surface of the ocean. The expectation was that it might take anywhere from a decade to half a century for the atmospheric temperature to get into what we refer to as 'equilibrium'. Equilibrium is the idea that if we froze the concentration of greenhouse gases in the atmosphere right now – at somewhere around 385 parts per million – and stabilized it there, it might take anywhere from 10 to 30, 40, or even 50 years for the air to stop getting warmer, because until the surface of the ocean is in equilibrium with the air temperature, it will keep getting warmer until it reaches an equilibrium.

Thirty years ago that was considered essentially the only role that oceans played. Now oceans are recognized as being active participants in the circulation of heat around the globe, and phenomena like the El Niño in the South Pacific Ocean are recognized to be the result of active oceanic participation in the circulation of heat around the globe.

Thirty years ago, clouds were ignored. Now it is recognized that clouds – depending on the size of the water droplets, on whether the clouds are ice crystals instead of water droplets, depending on whether the clouds are over the oceans or over land masses – can either be absorbers of outgoing radiation or reflectors of incoming radiation. None of these things was known, or at least was not appreciated 30 or even 20 years ago.

Part of what I am explaining is that this is a new subject. I participated, in the late 1970s, in two large energy studies. Recall that we had what was known as an energy crisis in the 1970s and early 1980s. I was in one project on nuclear energy: we had nuclear engineers, nuclear medical people, petroleum engineers and economists – all sorts of people. We published a 400-page book, '*Nuclear Power: Issues and Choices*' (Schelling and Nuclear Energy Policy Study Group, 1977). It had two pages on the greenhouse problem. It had a whole chapter on how bad coal might be for your lungs, but two pages out of 400 in a book about nuclear energy. If today you had a book on nuclear energy, one-quarter of it would be on nuclear weapons proliferation, one-quarter would be on how to dispose of nuclear waste, and half the book would be on the fact that nuclear power does not produce any greenhouse gases! Yet just 30 years ago, in 1977, a whole book had two pages mentioning the greenhouse problem.

I was in another project, called 'Energy, the next 20 years' (it is an embarrassing title now). We published a 600-page book in 1979. I recently looked in the index: there are 10 scattered references to the greenhouse problem, adding up to less than 10 pages out of a 600-page book. In the 20 years that followed, the greenhouse problem began to dominate all energy discussions. That is how new the subject is, and, while I sympathize with people who get impatient about how little is being done to mitigate the problem, this is, nevertheless, still early in the era of concern with the greenhouse problem.

I mentioned that there is uncertainty about how much global warming may result from various levels of greenhouse gases, and then there is uncertainty about how those translate into climates. Then you have to translate those climate changes into the impact on productivity, health, comfort, recreation and everything else. There I find one of the greatest difficulties is imagining the kind of world we shall be living in when these climate changes begin to be serious, 50 or 100 years from now. How do you come to grips with the way life is going to change?

One possible way is to look back at how much life has changed in the last 50 or 75 years. Sixty years ago, we did not have electronics, we did not have nuclear power or nuclear medicine, we had no antibiotics, hardly any vaccines; there were no plastics – we had celluloid.

If I ask myself: back in the 1920s, if people were predicting the kind of climate change that some people are predicting now for the next 80 years, how would people have reacted in the 1920s? I expect that they would be concerned about what was going to happen to mud. In the 1920s, mud was a serious environmental problem. Automobile tyres were about 2.5 inches in diameter, pumped up to 60 pounds per square inch, as hard as wood, and in mud automobiles got no traction. My uncle Harry used to make money from cars that stalled in the muddy road that went by his farm; he would take a team of horses and pull the cars out of the mud.

You could not ride a bicycle in mud, and in those days most children walked to school, and mud was a problem. People would have wondered 'What's going to happen in the summertime when the climate changes?', maybe not realizing that by the end of the twentieth century, at least in the United States, everything would be paved solid – my children have never driven a car in mud, let alone my grandchildren.

I also ask myself, since the main impact of climate change is likely to be on agriculture, what would a 90-year-old farmer respond, if you asked him, 'What has happened to farming in the 80 years or so that you've been working on a farm? What are the most dramatic changes?'. He would likely say, 'Well, the disappearance of the horse, the disappearance of kerosene lamps, the telephone; my car has a heater now'. I would say, 'What about climate change?', and he would reply, 'Well, it's hard to know what the impact on farm productivity has been, because we have different crops now. We never had soy beans in the 1920s, now we grow soy beans. We have hybrid corn, we have insecticides and herbicides and things of that sort, artificial fertilizer, and everything is so changed, I don't know what to attribute to any climate change'.

So much has changed in the last 80 years that it is important to recognize how much may change in the next 80 years. It is important not to imagine climate change being superimposed simply on life as we know it now.

We can be sure of a few things. One is that, in a developed country like France or the United States, not much economic production for the market is affected by climate. In the United States, automobiles can be assembled in any state. They are not assembled in Alaska, but that is because of distance rather than climate; open-heart surgery can be performed in any state; banking and insurance can be done – well or badly – in any state of the Union; pharmaceuticals can be produced in any state; television broadcasting can be done in any state.

Except for agriculture – farming, forestry, fisheries – there is not much that *is* affected by climate. In a country like France or the United States, despite the remarkable political power of farmers, farming is a small part of the economy. In the United States, farming is less than 2.5 per cent of its GDP. Some may say, 'But it's an important part: without farming you can't have food'. The way to look at it is: if the cost of producing food in a country like the United States would double or triple or quadruple over the next 80 years, it would probably mean that per capita income would double, not by the year 2060, but by the year 2063 or 2064. The impact of climate on GDP per capita in the developed countries is not likely to be large.

The situation is more acute in developing countries, where as much as half the population can depend on agriculture, much of it subsistence agriculture. So people in developing countries are vulnerable to climate change – not that all climate change will necessarily be adverse – but they are vulnerable in a way that Americans, Canadians, French, Australians or Israelis are not vulnerable. So the countries that are least able to do anything about greenhouse gases are the countries that have the most at stake, the most to lose from any drastic climate change.

There is a lot of concern about the effect of climate change on health. For reasons that I do not understand, most tropical vector-borne diseases, whether malaria or schistosomiasis or river blindness or whatever it may be, become more virulent when the temperature goes up, and most of those diseases borne by fleas, flies, gnats, mosquitoes and snails will spread territorially as the tropics expand with global warming. So the health impact has to be considered. But again we want to recognize that things are likely to change over the next 50 years.

A good example is Singapore and Malaysia. Singapore is separated from Malaysia by one kilometre of seawater – they have identical climates. Half a century ago, Singapore and Malaysia were part of the same nation. Since then, Singapore has developed to become about the highest-standard-of-living country in the world. Malaysia has also developed, but nothing like Singapore. There is no malaria to speak of in Singapore; there is a lot of malaria in Malaysia. Now that is partly because Singapore is a rich, compact state that is able to eradicate mosquitoes. If a Singaporean gets malaria, it is probably because he or she spent the weekend in Malaysia and got bitten by a mosquito and came back having contracted malaria. But the Singaporean is healthy to begin with.

Malaria kills a million people every year, but it does not kill healthy people; it kills people who are undernourished or suffering some debilitating disease already. The Singaporean who gets malaria in Malaysia receives excellent medical treatment. It is a serious illness, but it is not life-threatening.

The same with measles: measles kills a million children every year. When I was a boy, everybody got measles. You stayed away from school for a week or 10 days; it was not a terribly uncomfortable disease. How then does measles kill a million children every year? It is killing children who are weak to begin with, undernourished, sometimes hungry, but often just lacking the nutrients that would make them healthy.

If in the developing world, in the course of the next 50 years, we can reduce the incidence of childhood hunger and malnutrition, the impact of climate change on health, on longevity in the developing world, will be hugely attenuated. I mention all this to suggest that the best defence against climate change for the vulnerable parts of the world – meaning the so-called 'developing countries' – is going to be their own development, so they become less dependent on subsistence agriculture and less vulnerable to the various kinds of diseases that can afflict them if they are undernourished to begin with.

That leads me to a somewhat pessimistic conclusion about the political response to the threat of climate change. It is going to be hard to get people in a country like the United States to take climate change seriously, if they are made to believe that the vulnerable parts of the world are somewhere else. Al Gore probably exaggerated the threat to Americans, and I do not know how to get them to take it seriously unless we *do* exaggerate the threat to them, or else persuade them that peace and security in the long run depend on the successful development of the parts of the world that are still poor and vulnerable to climate change.

I want to say a little about mobilizing the nations of the world to do something about global warming. I know of no peacetime historical precedent for the kind of international cooperation that is going to be required to deal with climate change. In wartime, the Soviet Union, the United Kingdom and the United States energized themselves to defeat the Axis; nobody had to enforce the military commitments that were undertaken by the British, the Americans and the Soviets . . . I cannot think of anything in peacetime of the *magnitude* of the cooperation that is going to be required among the main nations of the world.

Two things are going to make any enforceable system difficult. First, I do not see any possibility that we could agree on what is the limit on the concentration of greenhouse gases in the atmosphere. Reducing emissions is a short-range objective. Eventually the problem will be to stabilize the concentration in the atmosphere, which means getting emissions down to where they are not adding to the concentration. The uncertainties are still at least as large as a factor of three, so to decide what the target is for the end of the twenty-first century is currently not feasible.

Second, I also see no possibility that we can have *enforceable* national limits on greenhouse gas emissions. People talk about 'binding commitments': well, commitments are supposed to be binding, that is what the word means, but who is going to enforce the United States' living within a certain quota? Who is going to enforce it on Guatemala, or Namibia? Nobody is going to propose military force; I doubt that anybody is going to threaten economic boycott. We are talking about what is inherently a voluntary system.

I was impressed a few years ago when both France and Germany were about to violate the fiscal provisions of the European Union. No nation was supposed to incur budget deficits in excess of 3 per cent of GDP for three years in a row. In 2004, France and Germany were both just about to complete their third year in a row with budget deficits in excess of 3 per cent of GDP – and nobody expected anything to happen, and nothing did happen. It is hard to imagine a greenhouse regime that would be tighter, more binding, more serious than the European Union, and if you cannot enforce a primary rule on the members of the European Union, I do not see how you are going to enforce carbon quotas on major countries.

My only historical example of cooperation on the scale that is likely to be required is the North Atlantic Treaty. Beginning in 1951, 15 nations negotiated over what they would do to build up defence forces against the possibility of attack from the East. Those nations did incur major commitments about drafting young men into military service and finding the budgetary means of equipping them, maybe letting them be stationed in Germany, finding real estate for manoeuvres, for barracks, for military pipelines and things of that sort. There was no enforcement. The NATO nations simply incurred serious commitments and for the most part lived up to them – at great expense. But there was no enforcement mechanism. They met their commitments because self-respecting nations incur commitments, and as best they can, they meet those commitments.

So I propose that the way to mobilize the world in relation to the greenhouse problem is first to get the United States on board, and that is almost certain to happen now. Then the developed countries, maybe for convenience the OECD, could fashion themselves a little on the North Atlantic Treaty and negotiate what they will do. Here it is important to note that in NATO, the commitments were not to results, but to what was *actually to be done*. In Kyoto, the commitments were to the *results* that were to be anticipated by 2010 or 2012. Most nations that committed themselves at Kyoto had no clear idea of what would be required in order to meet those commitments.

If there is to be a workable 'successor' to Kyoto it is important that the commitments to be undertaken by the major nations should be commitments to what they will *do*, not to what the results will be 15, 20 or 25 years down the road.

One of the first things to do is research and development. Right now, in both the United States and Western Europe, energy research and development is much lower than it was in the mid-1970s. It was taken seriously in the 1970s; apparently it is not taken that seriously now.

What kind of research and development? Well, one kind concerns capturing and sequestering carbon dioxide. It has been known for 50 years that you can pump carbon dioxide, as maybe produced at an electric power plant, down into an oil well. It has been used that way to help extract oil from wells that have been substantially depleted. So the technology of getting carbon dioxide underground has been around for 40 or 50 years, but it was in nobody's interest to find out what happened to the carbon dioxide 10, 20, 30 or 40 years later.

So there has been no research and development on where it may stay once it has been put there, or how it should be sealed in to prevent it from leaking out. We should engage in research and development on what kind of sequestration will work, bearing in mind that if you put it underground and it leaks out within 50 or 75 years, you have not accomplished anything.

The Chinese are building coal-fired electric power plants at the rate of more than one per week. It is crucially important, if sequestration is going to be workable, to find out where to locate those power plants. It is cheaper to transport electricity than to transport carbon dioxide. So plants should be built close to where the carbon dioxide can be sequestered, and the sooner it is ascertained where that can be done, the sooner it will be known where these 50 or 75 plants per year that are being built should be located.

This is going to require a lot of geological exploration and experimentation, and that should be getting attention in the United States, China, Russia, Germany, anywhere that is heavily dependent on coal. That is the kind of thing on which so little is being done. The United States had a pilot plant for sequestering carbon dioxide, and it was cancelled just a few months ago and nobody has given a good explanation of why it was cancelled, whether it was because there is a superior technology coming up soon, or whether it was becoming too expensive, or there just was not enough interest.

Let me just mention now what some of the most serious problems may be from climate change. One is the possibility that the average global surface atmospheric temperature might not stay within the range of 3, 4 or 5°C, but that what Weitzman calls 'the fat tail of the distribution' might lead to an increase of 10, 12 or 15°C. There are enormous methane deposits all around the world on the continental shelves at the bottom of the oceans, but essentially they are shallow oceans that are available to China, to Mexico, to the United States, almost anywhere in the world, and nobody knows what might cause those methane deposits to be released.

Methane is a potent greenhouse gas. It does not have a long residence time in the atmosphere, 10 or 15 years, but if enough of it is released, it could aggravate the greenhouse problem. It looks as if there may be lots of methane trapped underneath the permafrost in Siberia, Canada and Alaska, and some of that permafrost is beginning to soften and melt. Carbon dioxide is already being emitted from the melting permafrost, but the worry is that there may be lots of methane underneath that has been trapped in by the frozen surface, and when the surface becomes unfrozen the methane may leak out. The problem is that there may be a multiplier effect, because the more methane leaks out, the more temperature will rise, and therefore more methane released.

The second thing to worry about is a body of ice called the West Antarctic Ice Sheet, which is what they call 'grounded ice': it is resting on the bottom of the ocean, but it rises a kilometre or two above the ocean level. When the Arctic floating ice melts it does nothing to the sea level, but this grounded ice is like an iceberg so big that it is resting on the bottom, and if it should glaciate into the ocean, or break loose from where it is and collapse into the ocean, the estimate is that there is about 6 metres' worth of rise in ocean level in that particular body of ice.

When I was on a committee of the National Academy of Sciences in the early 1980s, we were worried about that, and we enlisted glaciologists to look into the question of how likely it is that the West Antarctic Ice Sheet will collapse and raise the ocean level. We were assured that it could happen, but not for two, three or four hundred years, and it would probably be gradual, and we would have plenty of warning.

By a strange coincidence, people became interested in Arctic and Antarctic ice just about the same time that we began to have satellite reconnaissance, and all of a sudden we can now measure the rate of flow of glaciers by the metre, not the kilometre. There is considerable concern now that warming of the water in Antarctica may lead this West Antarctic Ice Sheet to break loose . . . and 6 metres of ocean level is a lot: Copenhagen is under water; Stockholm is under water; Los Angeles is under water; the US President has to go by boat to the Capitol.

A lot of real estate can be saved with dykes, the way the Dutch have been doing for several hundred years, but you cannot save a country like Bangladesh, where tens of millions of people live within 20 feet of sea level. If enough dykes could be built to save Bangladesh from the ocean, they would all drown in fresh water, because you cannot get rivers to flow up and over the dyke. In Rotterdam, the dykes are against the rivers, not against the ocean, because the rivers are above or at sea level, and Rotterdam is below sea level. So this is one of the things to worry about.

A final word on geoengineering. Geoengineering essentially means, as it has come to be used in the language of greenhouse problems, somehow reflecting away part of the sunlight, increasing what is known as the Earth's 'albedo', the reflectivity of either the surface or the cloud levels or something of the sort. It is known that volcanoes that emit sulphur have a significant cooling effect on the Earth. One example is Mount Pinatubo, which erupted in the Philippines in the early 1990s, and which had a measurable effect on air and ocean temperatures, because it had a lot of sulphur in it.

So the proposal is, if we are changing the Earth's climate by putting something in the atmosphere that impedes outgoing radiation, why can we not offset that by putting something in the atmosphere that intercepts *incoming* radiation? Now, sulphur is not considered a healthful thing to put in the air. On the other hand, the estimates are that if appropriate 'sulphur aerosols', as they are called, could be released into the *stratosphere*, it would tend to stay up there for a year or so. Most of the sulphur that goes into the atmosphere in China or Germany or the United States has a residence time of about a week or 10 days, and then it comes down. If something could be introduced into the stratosphere that would last for a year or so, it would take a small amount of sulphur compared to the amount that is currently being put into the atmosphere all around the Earth.

Undoubtedly, if we had 50 years to experiment and get ready, we could find something more benign than sulphur to put in the stratosphere. When I used to talk about this 15 years ago, half the audience thought I was crazy and half thought I was dangerous. But this is beginning to come out of the closet. Geoengineering figured heavily on the agenda of the International Scientific Congress on Climate Change held in Copenhagen in March 2009. So this is beginning to get attention, and we are hearing more and more about it. This is perhaps to be thought of as a last resort, in case over the next 50 years we find that doing anything to mitigate climate change has become politically impossible. But even as a last resort, the sooner there is some experimentation to find out how to perform this feat of geoengineering, where to do it over the Earth and what to put in the stratosphere, the sooner we shall know whether this is likely to be a safeguard.

It has an acute disadvantage: one reason why people concerned with the environment do not like to talk about geoengineering is that if this is seen as a cheap and easy solution to the greenhouse problem, they worry that interest will be lost in developing the technology to reduce carbon dioxide emissions. That is a dilemma which is difficult to face. But if it does turn out that geoengineering becomes attractive, it will be helpful to have some experimentation to find out just what are likely to be the benefits of this kind of geoengineering and what may be some of the dangers.

One of the dangers is that this looks to be such a *cheap* way of offsetting global warming that it might be within the economic capability of the United States alone or of China alone or of some state alone to engage in this, and that might mean that it becomes a matter of international conflict.

But the sooner we discover whether this is going to be feasible, and what some of the disadvantages may be to watch out for, the better. To give some notion of magnitude, the amount of incoming sunlight that would have to be reflected away to offset a doubling of the concentration of greenhouse gases is around 1.5 per cent of the incoming sunlight. Nobody bathing on the shores of the Mediterranean would notice the difference, and probably most astronomers with their telescopes would hardly notice the difference. It is a tiny fraction of the incoming sunlight.

Anyway, more will be heard about that in the years to come.

REFERENCES

Schelling, T and Ford Foundation Study Group (1979), *Energy: The Next Twenty Years*, Cambridge, MA: Ballinger.
Schelling, T. and Nuclear Energy Policy Study Group (1977), *Nuclear Power: Issue and Choices*, Cambridge, MA: Ballinger.

COMMENTS: TOWARDS AN ENLIGHTENED FORM OF DOOMSAYING

Jean-Pierre Dupuy

Climate change is made up of immense uncertainties and some solid certainties. In the text that follows, I shall comment on both, as well as on action to be taken, from a philosophical vantage point on the one hand, and from the perspective of a contributor to the IPCC on the other.

- *Argument 1: Certainties.* Global warming due to the accumulation of greenhouse gases in the atmosphere is a reality. Its most proximate cause is human activity. As a consequence, calamitous climate change is a dire prospect that cannot be ruled out if substantial reductions in carbon emissions are not implemented without delay. It is a well-known fact that the planet as we know it will not endure if China, India and Brazil, for example, follow the same developmental path as that of Europe and the United States.
- *Argument 2: Uncertainties.* There remain large uncertainties regarding the future. The modelling of climate change, for instance, cannot tell us whether the temperature of the planet by 2100 will have increased by 2 or 6°C. It must be noted, however, that half of that uncertainty results from the uncertainty regarding the policy that will be implemented, which may vary from a strong determined action to a lack of measures to reduce greenhouse gas pollution. Another observation: we know that tipping points exist, beyond which the behaviour of the systems under consideration changes dramatically, even catastrophically. Take, for example, the possible collapse of the West Antarctic Ice Sheet. The problem is that, in most cases, we do not know where these tipping points are. We shall only discover them once we have passed them, that is, once it is too late.
- *Argument 3: Action (or rather inaction).* Although the previous two points have been known for some time, no serious action up to the challenge is being taken, whether on the part of governments, corporations or ordinary people. The Kyoto Protocol was arguably a much weaker measure than required, and it was trampled under foot by mighty America. Knowing about the impending threats is obviously not sufficient for prompting a significant change in behaviour. A striking leitmotiv in Thomas Schelling's contribution is the phrase, 'it has been known for [a century, 50 years, decades . . .]', or 'it has [long] been understood': although true, this knowledge has had no effect whatsoever as far as action is concerned.

Uncertainties and Action

I would like to focus my remarks here on the loop between uncertainty and action. It is a gross and misleading simplification to treat, as many experts do, the climate and the global ecosystem as if they were a physical dynamical system. Human actions influence the climate, and global warming is partly a result of human activity. The decisions that will be made or not may have a major impact on the evolution of the climate at the planetary level. Depending on whether or not humankind succeeds in mitigating future emissions of greenhouse gases and stabilizing their atmospheric concentration, major catastrophes will occur or be averted. An organization such as the IPCC would have no raison d'être otherwise. If many scientists and experts ponder over the determinants of climate change, it is not only out of a love for science and knowledge; rather, it is because they wish to exert an influence on the actions that will be taken by the politicians and, beyond them, by the people themselves. The experts see themselves as capable of changing, if not directly the climate, at least the climate of opinion.

These observations may sound trivial. It is all the more striking that they are not taken into account, most of the time, when it comes to anticipating the evolution of the climate. When they are, it is in the manner of control theory: human decision is treated as a parameter, an independent or exogenous variable, and not as an endogenous variable. A crucial causal link is missing: the motivational link. What one decides to do depends in part on his or her representation of the future and, in particular, on the uncertainty that surrounds it. One might, for instance, surmise that the large uncertainties emphasized by Schelling have a paralysing effect on taking action. It is clear, then, that the decisions that will be made will depend, at least in part, on the kind of anticipation of the future of the system that will be made, and made public. This future will depend, in turn, on the decisions that will be made. A causal loop thus appears, which prohibits us from treating human action as an independent variable.

Being an Extremist

Schelling contrasts two forms of what I would like to call 'extremism', contending that neither makes any sense, economic or otherwise. He advocates the usual economic approach: 'Weigh the costs, the benefits, and the probabilities as best all three are known, and don't be obsessed with either extreme tail of the distribution'.

I disagree. I favour one position. Schelling defines the first form of extremism in the following terms: 'do nothing until we are absolutely sure the alternative is dangerous'; and the second, which he associates with the precautionary principle: 'do nothing until we are absolutely sure it's safe'. In Europe, everybody reacted to a glaring paradox: the paradox of the Bush administration ignoring the precautionary principle in environmental and public health issues, where precisely it first belongs, and resorting implicitly to it in the Iraqi crisis, with the notion of a pre-emptive strike. The burden of proof rested on Saddam Hussein's shoulders (Tarek Aziz, maybe influenced by his reading of Karl Popper, said a few weeks before the war started: 'how can one prove an absence?'). President Bush argued that the risk of weapons of mass destruction was great enough to warrant an attack, without absolute proof that Iraq was hiding such weapons. In the case of global warming, on the other hand, he maintained that much more research was needed before taking action. Schelling's chapter also refers to such inconsistencies.

I want to defend the (possibly) extremist view that we should focus on the worst-case scenario when two conditions are met: (i) the uncertainty is radical, which means in particular that resorting to some kind of cost–benefit analysis is infeasible, but not for de facto reasons (for example, our lack of knowledge regarding costs, benefits and probabilities), but for de jure reasons: conceptually, those notions make no sense whatsoever – we are beyond the realm of validity of economic analysis; and (ii) the stakes are immense: the very future of humankind is in jeopardy. The paradox here is that I would like to argue my point through a radical critique of the precautionary principle. In my mind, the precautionary principle is just a devious and baroque form of cost–benefit analysis. If you prefer, it is not extremist enough. My criticism is not of the position that the precautionary principle takes, but that it is conceptually vacuous.

A Critique of the Precautionary Principle

Let us recall the definition of the precautionary principle formulated in the Maastricht Treaty: 'The absence of certainties, given the current state of scientific and technological knowledge, must not delay the adoption of effective and proportionate preventive measures aimed at forestalling a risk of grave and irreversible damage to the environment at an economically acceptable cost'. This text is torn between the logic of economic calculation and the awareness that the context of decision making has radically changed. On one side, there are the familiar and reassuring notions of effectiveness, commensurability and reasonable cost; on the other, there is the uncertain state of knowledge and the gravity and irreversibility of damage. It would be all too easy to point out that if uncertainty prevails: no one can say what would be a measure proportionate (by what coefficient?) to a damage that is unknown; one therefore cannot say if the damage will be grave or irreversible; no one can evaluate what adequate prevention would cost; and no one can say – supposing that this cost turns out to be 'unacceptable' – how one should go about choosing between the health of the economy and the prevention of the catastrophe.

Rather than belabour these points, I shall present three fundamental reasons why the notion of precaution is an ersatz good idea that belongs in cold storage. I shall try at the same time to understand why the need was felt, one fine day, to saddle the familiar notion of prevention with an upstart sidekick, precaution. Why is it that in the present situation of risks and threats, prevention is no longer enough?

The first serious deficiency that hamstrings the notion of precaution is that it does not properly gauge the type of uncertainty with which we are confronted at present.

The French official report on the precautionary principle (Kourilsky and Viney, 2000) introduces what initially appears to be an interesting distinction between two types of risks: 'known' risks and 'potential' risks. It is on this distinction that the difference between prevention and precaution is made to rest: precaution would be to potential risks what prevention is to known risks.

A closer look at the report in question reveals: (i) that the expression 'potential risk' is poorly chosen, and that what it designates is not a risk waiting to be realized, but a hypothetical risk, one that is only a matter of conjecture; and (ii) that the distinction between known risks and hypothetical risks (the term I shall adopt here) corresponds to an old standby of economic thought, the distinction that John Maynard Keynes and Frank Knight independently proposed in 1921 between risk and uncertainty. A risk can in principle be quantified in terms of objective probabilities based on observable frequencies; when such quantification is not possible, one enters the realm of uncertainty.

The problem is that economic thought and the decision theory that underlies it were destined to abandon this distinction as of the 1950s in the wake of Leonard Savage's successful introduction of the concept of subjective probability and the corresponding philosophy of choice under conditions of uncertainty: Bayesianism. In Savage's axiomatic, probabilities no longer correspond to any sort of regularity found in nature, but simply to the coherence displayed by a given agent's choices. In philosophical language, every uncertainty is treated as an epistemic uncertainty, meaning an uncertainty associated with the agent's state of knowledge. It is easy to see that the introduction of subjective probabilities erases the distinction between uncertainty and risk, between the risk of risk and risk, between precaution and prevention. If a probability is unknown, a probability distribution is assigned to it 'subjectively'. Then the probabilities are composed following the computation rules of the same name. No difference remains compared to the case where objective probabilities are available from the outset.

Uncertainty owing to lack of knowledge is brought down to the same level as intrinsic uncertainty due to the random nature of the event under consideration. A risk economist and an insurance theorist do not see and cannot see any essential difference between prevention and precaution and, indeed, reduce the latter to the former. In truth, one observes that applications of the 'precautionary principle' generally boil down to little more than a glorified version of 'cost–benefit' analysis.

Against the prevailing 'economism', it is urgent to safeguard the idea that everything is not epistemic uncertainty. One could, however, argue from a philosophical standpoint that such is really the case. The fall of a die has supplied most of our languages with the words for chance or accident. The fall of a die is a physical phenomenon that is viewed today as a low-stability deterministic system, sensitive to initial conditions, and therefore unpredictable – a 'deterministic chaos', in current parlance. But an omniscient being – the God of whom Pierre-Simon Laplace did not judge it necessary to postulate the existence – would be able to predict on which side the die is going to fall. Could one not then say that what is uncertain for us, but not for this mathematician-God, is uncertain only because of lack of knowledge on our part? And therefore that this uncertainty, too, is epistemic and subjective?

The correct conclusion is of a different nature. If a random occurrence is unpredictable for us, this is not because of a lack of knowledge that could be overcome by more extensive research; it is because only an infinite calculator could predict a future which, given our finiteness, we shall forever be unable to anticipate. Our finiteness obviously cannot be placed on the same level as the state of our knowledge. The former is an unalterable aspect of the human condition; the latter, a contingent fact, which could at any moment be different from what it is. We are therefore right to treat the random event's uncertainty for us as an objective uncertainty, even though this uncertainty would vanish for an infinite observer. Now, our situation with respect to new threats is also one of objective, and not epistemic, uncertainty. The novel feature this time is that we are not dealing with a random occurrence either, for each of the catastrophes that hovers threateningly over our future must be treated as a singular event. Neither random, nor epistemically uncertain, the type of 'risk' that we are confronting is a monster from the standpoint of classic distinctions. It merits special treatment, which the precautionary principle is incapable of providing.

Three arguments seem to me to justify the assertion that the uncertainty here is not epistemic, but anchored in the objectivity of the relationship binding us to phenomena.

1. The first argument has to do with the complexity of ecosystems. This complexity gives them an extraordinary robustness, but also, paradoxically, a high vulnerability. They can hold their own against all sorts of aggressions and find ways of adapting to maintain their stability. This is only true up to a point, however. Beyond certain critical thresholds, ecosystems veer abruptly into something different; this can be in the form of phase changes of matter, complete collapse or formation of other types of systems with properties that are potentially highly undesirable for people. In mathematics, such discontinuities or tipping points are called catastrophes.

   This sudden loss of resilience gives ecosystems a particularity that no engineer could transpose into an artificial system without being immediately fired from his or her job: the alarm signals go off only when it is too late. As long as the thresholds remain distant, ecosystems may be handled with impunity. In this case, cost–benefit analysis appears useless, or bound to produce a result known in advance, since there seems to be nothing to weigh down the cost side of the scales. That is why humanity was able to blithely ignore, for centuries, the impact of its mode of development on the environment. Cost–benefit analysis becomes meaningless, however, the closer one gets to the critical thresholds, which must not be crossed at any cost. It is here that we see how useless economic calculation is, for the reasons lie in the objective structural properties of ecosystems and not in a temporary insufficiency of our knowledge.

2. The second argument concerns systems created by humans, let us say technical systems, which can interact with ecosystems to form systems of a hybrid nature. Technical systems display properties quite different from those of ecosystems. This is a consequence of the important role that positive feedback loops play in them. Small fluctuations early in the life of a system can end up being amplified, giving it a direction that is perfectly contingent and perhaps catastrophic, but which, from the inside, assumes the lineaments of fate. This type of dynamic or history is obviously impossible to foresee. In this case as well, the lack of knowledge does not result from a state of things that could be changed, but from a structural property. The non-predictability is fundamental.

3. Uncertainty about the future is equally fundamental for a third reason, logical this time. Any prediction regarding a future state of things that depends on future knowledge is impossible for the simple reason that to anticipate this knowledge would be to render it present and would

dislodge it from its niche in the future. The most striking illustration is the impossibility of foreseeing when a financial bubble will burst. This incapacity is not due to a shortcoming of economic analysis, but to the very nature of the speculative phenomenon. Logic is responsible for the incapacity, and not the insufficient state of knowledge or information. If the collapse of the speculative bubble or, more generally, the onset of a financial crisis were anticipated, the event would occur at the very moment that it was anticipated and not at the predicted date. Any prediction on the subject would make it invalid at the very moment it was made public.

When the precautionary principle states that the 'absence of certainties, given the current state of scientific and technical knowledge, must not delay', and so on, it is clear that it places itself from the outset within the framework of epistemic uncertainty. The presupposition is that we know we are in a situation of uncertainty. It is an axiom of epistemic logic that if I do not know p, then I know that I do not know p. Yet, as soon as we depart from this framework, we must entertain the possibility that we do not know that we do not know something. An analogous situation obtains in the realm of perception, with the blind spot, that area of the retina unserved by the optic nerve. At the very centre of our field of vision, we do not see, but our brain behaves in such a way that we do not see that we do not see. In cases where the uncertainty is such that the uncertainty itself is uncertain, it is impossible to know whether or not the conditions for the application of the precautionary principle have been met. If we apply the principle to itself, it will invalidate itself before our eyes.

Moreover, 'given the current state of scientific and technical knowledge' implies that a scientific research effort could overcome the uncertainty in question, whose existence is viewed as purely contingent. It is a safe bet that a 'precautionary policy' will inevitably include the edict that research efforts must be pursued – as if the gap between what is known and what needs to be known could be filled by a supplementary effort on the part of the knowing subject. It is not uncommon, however, to encounter cases in which the progress of knowledge comports an increase in uncertainty for the decision maker, something which is inconceivable within the framework of epistemic uncertainty. Sometimes, to learn more is to discover hidden complexities that make us realize that the mastery we thought we had over phenomena was in part illusory.

The second serious deficiency of the precautionary principle is that, unable to depart from the normativity proper to the calculus of probabilities, it fails to capture what constitutes the essence of ethical normativity concerning choice in a situation of uncertainty. Here, I am referring to the concept of 'moral luck' in moral philosophy. I shall introduce it with the help of two contrasting thought experiments.

In the first, one must reach into an urn containing an indefinite number of balls and pull one out at random. Two-thirds of the balls are black and only one-third are white. The idea is to bet on the colour of the ball before seeing it. Obviously, one should bet on black. And if one pulls out another ball (after replacing the first one into the urn), one should bet on black again. In fact, one should always bet on black, even though one foresees that one out of three times on average this will be an incorrect guess. Suppose that a white ball comes out, so that one discovers that the guess was incorrect. Does this *a posteriori* discovery justify a retrospective change of mind about the rationality of the bet that one made? No, of course not; one was right to choose black, even if the next ball to come out happened to be white. Where probabilities are concerned, the information as it becomes available can have no conceivable retroactive impact on one's judgement regarding the rationality of a past decision made in the face of an uncertain or risky future. This is a limitation of probabilistic judgement that has no equivalent in the case of moral judgement.

In the second example, a man spends the evening at a cocktail party. Fully aware that he has drunk more than is wise, he nevertheless decides to drive his car home. It is raining, the road is wet, the light turns red, and he slams on the brakes, but a little too late: after briefly skidding, the car comes to a halt just past the pedestrian crosswalk. Two scenarios are possible: either there was nobody on the crosswalk, and the man has escaped with no more than a retrospective fright; or the man ran over and killed a child. The judgement of the law will not be the same in both cases. Here is a variant: the man was sober when he drove his car. He has nothing for which to reproach himself. But there is a child whom he runs over and kills, or else there is not. Once more, the unpredictable outcome will have a retroactive impact on the way the man's conduct is judged by others and also by the man himself.

Here is a more complex example devised by the British philosopher Bernard Williams (1981), which I will simplify. A painter – we'll call him 'Gauguin' for convenience – decides to leave his wife and children and go to Tahiti to live a different life, one in which he hopes will allow him to paint the masterpieces that it is his ambition to create. Is he right to do so? Is it moral to do so? Williams defends with great subtlety the thesis that any possible justification of his action can only be retrospective. Only the success or failure of his venture will make it possible for us – and him – to cast judgement. Yet whether Gauguin becomes a painter of genius or not is in part a matter of luck – the luck of being able to become what one hopes to be. When Gauguin makes his painful decision, he cannot know what, as the saying goes, the future holds in store for him. To say that he is making a bet would be incredibly reductive. With its appearance of paradox, the concept of 'moral luck' provides just what we were missing in order to describe what is at stake in this type of decision made under conditions of uncertainty.

Like Williams's Gauguin, but on an entirely different scale, humanity, taken as a collective subject, has made a choice in the development of its potential capabilities, which brings it under the jurisdiction of moral luck. It may be that its choice will lead to great and irreversible catastrophes; it may be that it will find the means to avert them, to get around them, or to get past them. No one can tell which way it will go. The judgement can only be retrospective. It is possible, however, to anticipate, not the judgement itself, but the fact that it must depend on what will be known once the 'veil of ignorance' cloaking the future is lifted. Thus, there is still time to ensure that our descendants will never be able to say 'too late!' – a 'too late' that would mean that they find themselves in a situation where no human life worthy of the name is possible.

The most important reason that leads me to reject the precautionary principle is still to come. It is that, by placing the emphasis on scientific uncertainty, it utterly misconstrues the nature of the obstacle that keeps us from acting in the face of catastrophe. The obstacle is not uncertainty, scientific or otherwise; the obstacle is the impossibility of believing that the worst could happen.

Let us pose the simple question as to what the practice of those who govern us was before the idea of precaution arose. Did they institute policies of prevention, the kind of prevention with respect to which precaution is supposed to innovate? Not at all. They simply waited for the catastrophe to occur before taking action – as if its coming into existence constituted the sole factual basis on which it could be legitimately foreseen – too late of course.

Even when it is known that it is going to take place, a catastrophe is not credible: that is the principal obstacle. On the basis of numerous examples, an English researcher identified what he called an 'inverse principle of risk evaluation': the propensity of a community to recognize the existence of a risk seems to be determined by the extent to which it thinks that solutions exist. To call into question what we have learned to view as progress would have such phenomenal repercussions that we do not believe we are facing catastrophe. There is no uncertainty here, or very little. It is at most an alibi.

In addition to psychology, the question of future catastrophe brings into play a whole metaphysics of temporality. The world experienced the tragedy of 11 September 2001 less as the introduction into reality of something senseless – and therefore impossible – than as the sudden transformation of an impossibility into a possibility. The worst horror has now become possible, one sometimes heard it said. If it has become possible, then it was not possible before. And yet, common sense objects: if it happened, then it must have been possible.

Henri Bergson describes what he felt on 4 August 1914 when he learned that Germany had declared war on France:

> In spite of my shock, and my belief that a war would be a catastrophe even in the case of victory, I felt . . . a kind of admiration for the ease with which the shift from the abstract to the concrete had taken place: who would have thought that so awe-inspiring an eventuality could make its entrance into the real with so little fuss? This impression of simplicity outweighed everything. (Bergson, 1991, pp. 1110–11, our translation)

Now, this uncanny familiarity contrasted sharply with the feelings that prevailed before the catastrophe. War then appeared to Bergson at one and the same time as probable and as impossible: a complex and contradictory idea, which persisted right up to the fateful date.

In reality, Bergson deftly untangles this apparent contradiction. The explanation comes when he reflects on the work of art: 'I believe it will ultimately be thought obvious that the artist creates the possible at the same time as the real when he brings his work into being', (ibid., p. 1340). One hesitates to extend this reflection to the work of destruction. And yet, it is also possible to say of the terrorists that they created the possible at the same time as the real.

Catastrophes are characterized by this temporality that is in some sense inverted. As an event bursting forth out of nothing, the catastrophe becomes possible only by 'possibilizing' itself (to speak in the manner of Jean-Paul Sartre who, on this point, learned the lesson of his teacher Bergson well). And that is precisely the source of our problem. For if one is to prevent a catastrophe, one needs to believe in its possibility before it occurs. If, on the other hand, one succeeds in preventing it, its non-realization holds it in the realm of the impossible, and as a result, the prevention efforts will appear useless in retrospect.

REFERENCES

Bergson, H. (1991), *Œuvres*, Édition du centenaire, Paris: PUF.
Kourilsky, P. and G. Viney (2000), *Le principe de précaution*, Report to the Prime Minister, Paris: Odile Jacob.
Williams, Bernard (1981), *Moral Luck*, Cambridge: Cambridge University Press.

2. Economics in the environmental crisis: part of the solution or part of the problem?

Olivier Godard

As a positive, that is to say explanatory, science, economics must analyse the behaviour of agents enjoying a certain degree of freedom but subject to the constraints imposed on them by nature and institutions.

As a normative science, economics must investigate the best way to organise production, distribution and consumption.

(Malinvaud, 1975, p. 2)

Thinking like a mountain.

(Leopold, 1949 [1989], p. 129)

INTRODUCTION

The Earth's climate is on the brink of serious disruption. According to the vast majority of competent scientists, and in particular the assessment reports regularly published since 1990 by the Intergovernmental Panel on Climate Change (IPCC), this is due to greenhouse gas (GHG) emissions produced by the human activities of production, transport and consumption. This can be traced back at least as far as the nineteenth century, when the industrial revolution started to spread through Europe, and which today has reached most regions of the world – notably Asia – with the exception of Africa. Global warming is the biggest environmental externality ever generated by the human economy, to borrow a phrase from Nicholas Stern, main author of the enormously influential Stern Review, a report on the economics of climate change commissioned by the British government (Stern, 2006).

Economic activity – given its various forms and physical content over the last 150 years – lies, without a doubt, at the heart of the climate risk. But what about economic thinking? To what extent do the concepts and methods developed by the discipline of economics, particularly during the last 50 years, provide the appropriate analytical tools for understanding

the situation, or the best normative benchmarks for determining what
action to take? The case for the prosecution is easily made. Is this discipline not largely responsible for the situation? After all, if the real economy
has failed us so badly by endangering the global environmental conditions
of human life, is this not because the discipline of economics has turned
out to be a bad adviser or, worse, blind to what was at stake, because of
its truncated conceptual foundations? Should we not, for example, call
into question the obsession with economic growth that many economists
share,[1] or the confidence that many of them have in the aptitude of real
markets to accurately reveal the preferences of individuals,[2] or the premise
that the collective good we should aim for is only the one that maximizes
the level of satisfaction of the preferences of consumers?

These are all legitimate questions, but we must not draw hasty conclusions
from them for at least two reasons. First, whatever influence economists
have, the real economy is not the direct transcription of their advice and recommendations. Second, and most importantly, the discipline of economics
is not homogeneous enough to produce a unique set of principles and rules
for organizing economic activity. On the contrary, economics is riddled
with oppositions between diverging theses and schools of thought to such
an extent that one observer of the running of public affairs concluded that if
economic language appears to have taken over centre stage in public policy
making, the hard decisions made behind the scenes are still governed by
political considerations (see Majone, 1989): the multiplicity of schools and
the controversies between economists enables public policy makers to use
economic rhetoric as and when they feel necessary without being captive to
the general framework or to the recommendations of any particular economist. They can simply choose, out of the range of proposals available, those
that are best suited to their programme and interests. Nevertheless, this
relative distance between the real economy and economic representations
of the world is not wide enough to exonerate the essential elements of the
theoretical corpus of this discipline without further examination.

Sticking to the issue of climate change, I therefore invite the reader to a
wider reflection, not only on the relations between real economic activity
and the natural environment – of which climatic conditions are an absolutely critical component – but also on the representation of environmental
issues made by the discipline of economics. I shall bring these two themes
of reflection together by exploring the effects on the natural environment
that are logically generated by the prevailing economic representation of
the world. For that purpose, I shall consider the patterns of thinking that
constitute the economic view of the world as determinants of individual
and collective human action and, in consequence, as operators producing real effects on the world. In other words, if the real economy *was* the

simple transcription of the concepts and tools of economics, what would be the real impact on the world? Would we discover the solid foundations of sustainable development that (almost) everybody agrees to be the way forward in the twenty-first century, to make sure that 'future generations have a future', to borrow from the title of a book by Christian de Perthuis (2003), or would we, on the contrary, discover the 'seeds of disaster', to borrow from another book title (Auerswald et al., 2006)?

The guiding thread of this chapter, which will become clear when the reader reaches the end of the fifth section, is to consider the apparently heterogeneous group of elements that make up the current landscape of environmental economics as a complex structure that links self-referential and hetero-referential viewpoints, without being limited to one or the other. Only a full deployment of this structure can render the environmental stakes fully intelligible. This makes the field of environmental socio-economics – still in its first stages – the pivotal key to understanding the environmental crisis, and in particular, global climate disruption.

LANDMARKS IN THE HISTORY OF ENVIRONMENTAL ECONOMIC THOUGHT

If I had to summarize the history of environmental economic thought, I would start by highlighting the ambition to discipline this field using the most general economic concepts and models, created in other contexts to resolve other problems, such as price formation, market imperfections, the dynamics of growth, the financing of public expenditure, the economics of information, asset evaluation and pricing, investment choices, insurance risk coverage, the formation and distribution of rents, and so on. Immediately afterwards, I would mention the recurrent, lasting unease, sometimes leading to intellectual and moral revolt against the reductionism of that approach, which sees environmental problems as nothing more than classic economic problems of restoring optimality in the allocation of goods between agents with incompatible demands.

The emergence of the environment as a political issue during the 1960s was indeed accompanied by criticism of economics in its most aggressive manifestations: economic growth by every means possible; the race for private profits, echoing the slogan 'private vices, public benefits' that Bernard de Mandeville exalted as a law of society at the beginning of the eighteenth century; the commodification of every aspect of social life; and the insatiable desire for indiscriminate technical innovation as an imperative for stimulating demand. Economists also seized upon the new issue, most of them endeavouring to show that economics, as a science of scarcity, offers indispensable keys to understanding and brings to light essential principles of organization, concluding that it was important not to pick the wrong target or to throw the baby out with the bathwater.

Despite these often welcome contributions in defence of economics, misunderstandings and disagreements have persisted over the fundamental premises of modern economics, often veiled behind sophisticated mathematical models. And it is these premises that sometimes lead to a complete rejection of economic approaches. It is true that symmetrically, economists often consider that their analyses and recommendations have a kind of natural monopoly on reason, forgetting the foundations of their representations along the way; whence their propensity to cast any criticism of economics into the hell of irrationality.

Economic analysis has only really been confronted with the question of the environment since the 1960s, although the theoretical bases had been laid long before. On the phenomena of pollution, economists from Arthur Pigou (1920) to Ronald Coase (1960) introduced the idea of a gap between social costs and private costs resulting from the presence of externalities. On the exploitation and management of natural resources, Harold Hotelling (1931) laid the foundations for the economics of exhaustible resources, and Howard Scott Gordon (1954) and Milner Bailey Shaefer (1955) for renewable resources such as fisheries. We could go further back to the issues and debates of the nineteenth century on the source of value and land rent (Ricardo, 1817) or the growing imbalance between geometric population growth and the arithmetic growth in resources (Malthus, 1803). In the 1950s, Paul Samuelson (1954) formalized the concept of public goods from the point of view of public finance, and Tibor Scitovsky (1955) introduced and defined the distinctions between pecuniary and technological externalities. There was also the far-sighted book by Karl Kapp (1950) on the social costs of the market economy, showing the implications of company behaviour on the formation of the welfare state in this type of economy: internalizing profits and externalizing costs towards other people or the government, whether these costs take the form of pollution or the training and management of human resources.

A new cycle of interest in the threats and damage to the natural environment started to pick up steam in the mid-1960s among both economists and society at large. In the field of environmental economics, I would particularly mention the article by Burton Weisbrod (1964) introducing, for the first time, the idea of an option price not revealed by the market in a debate on the management of national parks in the United States, Ezra Mishan's book (1967) on the costs of economic growth, and Alan Kneese's book (1964) on the regional management of water quality.

At the end of the 1960s, the emergence of environmental problems on the political scene and the first reports on the 'damages of progress' (Farvar and Milton, 1972) revived fierce debate about the mid- and long-term antagonism or compatibility between the pursuit of economic growth, on the one hand, and the limited availability of natural resources and protection of the natural environment on a worldwide scale, on the other. The study of the question was systematized by the Club of Rome by using the formalism of system dynamics developed by Jay Forrester (1971) at the Massachusetts Institute of Technology. This led to publication of the report *The Limits to Growth*, which fuelled considerable public debate in Europe (Meadows et al., 1972), but which was subjected to sharp criticism from most economists for having ignored the role of economic mechanisms of regulation in its projection of imbalances.

In this context, economists adopted two different approaches to environmental issues. The first consisted in applying the conceptual and methodological resources already developed in the general corpus of welfare economics. The most self-confident representative of this approach – constantly advocating the relevance and superiority of its theoretical and normative foundations – was probably the English economist Wilfred Beckerman (1974), who taught at Oxford. This approach did not exclude certain incremental conceptual innovations within the standard theoretical framework, such as the theory of option values under uncertainty and irreversibility proposed simultaneously in 1974 by Claude Henry (1974) and by Kenneth Arrow and Anthony Fisher (1974). The second approach called for a profound theoretical rethinking of the way economic activity is represented in response to a historically unprecedented challenge. In the 1970s, this was the ambition of the programmes of bioeconomics or ecological economics, proposed by René Passet (1979) in France and Robert Costanza (1991) in the United States.

At the same time, there was growing disillusionment with the ideology – championed by Walter Rostow in his fight against communism and supported by the US superpower and like-minded international institutions such as the OECD and the World Bank – based on a supposed law of the stages of economic growth that all countries should follow (Rostow, 1960). This doctrine exalted economic growth as a superior good that – through diffusion and trickle-down effects – could solve not only all the problems of poverty and underdevelopment, but also those of the environment. Regarding environmental problems, a contemporary grandchild of this ideology has come to be known as the hackneyed concept of the 'environmental Kuznets curve'.[3] According to these views, a unique institutional and economic roadmap is to be followed by every country on the planet, and the maximization of growth is seen as the best long-term vector for the protection of the environment.

In this debate about economic growth, some of its first theorists, such as Robert Solow (1974, 1993), and others who followed, sought to reassure by counting on technical progress and the extended relation of substitutability between natural assets and reproducible goods within a utilitarian approach to welfare. One of the key moments was the 1974 Symposium of the *Review of Economic Studies* on the economics of exhaustible resources. On the other side, a powerful movement emerged, critical of both the foundations and hypotheses of the dominant neoclassical economic conception and also of its methods. Echoing Kenneth Boulding's 1966 declaration that society had already entered the age of the 'spaceship economy', it was in 1972, as mentioned above, that the first report to the Club of Rome on the limits to growth broke with the standard economic approach to these problems, both in its tools and its message. Furthermore, 1973 saw the promotion of the eco-development approach by Maurice Strong, Secretary-General of the UN Conference on the Human Environment convened in Stockholm in 1972 and Ignacy Sachs (1972, 1974, 1980), defining the conditions of new development strategies that take the new environmental context into consideration. According to these authors, the objectives of growth – considered still to be necessary for the majority of people – need to be harmonized not only with the specific natural, cultural and economic conditions of each region – and especially the rural zones of southern countries – but also with the objective of conserving the essential elements of the environment for future generations.

That same period also saw the attempt by Nicholas Georgescu-Roegen to rethink economic activity on the basis of thermodynamics, going so far as to transpose the concept of entropy to the management of flows of matter circulated and transformed by industry. Here, the movement for degrowth found its most serious scientific support. Lastly, this period saw the birth of a new philosophical approach to nature, called 'environmental ethics' (see Shrader-Frechette, 1981; Sagoff, 1988; Larrère, 1997; Afeissa, 2007), fuelled by deep dissatisfaction with the anthropocentric ethics of the utilitarianism on which welfare economics was founded. Here, I cite Richard Sylvan[4] (1973), Arne Naess[5] (1973) and John Baird Callicott[6] (1979).

Since then, the diversity of approaches has marked the field of environmental economics at least as much as it has marked the discipline of economics as a whole. All the schools that make up the latter (neoclassicism, neo-Marxism, neo- or post-Keynesianism, the Austrian tradition, socio-economics and so on) can also be found in the former – with a different audience, of course – adding their own differences and problems of dialogue to the divergent attitudes advocating either the application of an existing corpus or the constitution of a fresh approach.

A COHERENT TOPOLOGY OF THE FRAGMENTED DOMAIN OF ENVIRONMENTAL ECONOMICS

The field of environmental economics is so diverse that it is difficult to grasp its scientific unity. Should we limit ourselves to this observation of heterogeneity, putting it down to a lack of scientific maturity despite the decades of development? Or can we reach beyond the apparent diversity and discern a deeper unity linking rival scientific currents that are only too keen to consider themselves as being exclusive of the others? If we choose the second answer, then where can we find this unity and how can we characterize it?

Three Main Currents of Unequal Influence

The idea that I would like to propose to the reader is this: beyond the relations of rivalry between different schools, the field of environmental economics has, in a manner non-contingent upon the subject of its study, become organized around three main currents. The first is the neoclassical current. It treats the environment as a collection of goods that, despite certain particular conditions, fall within the general issue of the allocation of goods according to the preferences of individual agents. The second is 'ecological economics', which, at the limit, studies the human economy as it would study an ecological system, through, for example, the examination of energy transfer. This physics-centred approach may be balanced, however, by its tenants who adopt the most critical look or radical standards of fairness on policy issues. The third is socio-economics, focused on the interrelation between resources and environmental usages, institutions and social values and standards. For the last approach, institutions and social values are necessary mediations in the relationship between economy and nature, opposing any mechanistic views on the development of environmental problems as direct expressions of a new physical scarcity of nature.

The programme of neoclassical environmental economics is to apply concepts and methods drawn from the school's standard paradigm to environmental subjects (see Cropper and Oates, 1992; Oates, 1994; Sterner and van den Bergh, 1998): economics is conceived as a universe in itself, existing independently of social institutions but capable of being affected by them; methodological and normative individualism reduces collective phenomena to individual rationales; individual behaviour is explained in terms of expectations, rational choices and preferences; central importance is attached to the idea of equilibrium; intellectual priority is given to coordination through prices and contracts. In that respect, the environment has provided a new opportunity for expanding the fields of application of this school, which has, at the same time, set out to conquer other territories, such as the economics of crime and marriage or the economics of regulation and law.

Although subject to severe criticism, or even complete rejection, in the 1970s from such pioneers as Karl Kapp, René Passet and Ignacy Sachs, the neoclassical current has managed to impose itself, 30 years later, through the power of its formalizations and the weight of researchers using its conceptual framework. In the eyes of many, this manner of approaching the subject, whatever that may be, is the signature of the professional economist and is obviously *de rigueur* whenever scarcity is an issue. The theoretical corpus thus developed has become dominant notably in international institutions (OECD, World Bank and so on) and national administrations.

The second current aims not only to grasp the new issues and requirements generated by the question of the environment, but to rethink both the foundations and tools of economics from this perspective. 'Bioeconomics', 'ecological economics' or 'physico-economics' were the principal names given to this movement: the question of the environment was considered so decisive for the future of humankind that it was necessary to lay the foundations of a new science to do it justice. The subject assigned to this new science was the study of complex interactions between the human economy and the physical and biological functioning of the planet Earth on which that economy depends.

Initially promoted by relatively isolated scientific personalities from the community of economists, such as Kenneth Boulding (1966), Herman Daly (1968), Robert Ayres (Ayres and Kneese, 1969; Ayres, 1994), Nicholas Georgescu-Roegen (1971), René Passet (1979) and Malte Faber (Faber et al., 1987), this second approach sought the conceptual and methodological renewal of economics through an interdisciplinary approach in conjunction with the natural sciences developed during the twentieth century: thermodynamics, information theory, ecology, integrative biology – all sciences that take systems theory as their general frame of reference. The tools of this new science were to be borrowed both from the natural sciences and from those economic works most concerned with the functioning of the material economy, such as eco-energetics or material balance analysis, which have recently produced widely used instruments of evaluation, such as 'life-cycle analysis' or 'ecological footprint'. This resulted in a current that identifies itself and is identified by others through the formation of the International Society for Ecological Economics and its internationally reputed journal *Ecological Economics*.

The participants in this new, relatively heterogeneous current are sometimes guilty, however, of a form of reductionism symmetrical to that of which they accuse the neoclassical mainstream. The goal of achieving a better match with the conceptions of the natural sciences has proved to be so absorbing that some of the researchers in question have overlooked what it is that makes human society neither a natural system – although it does have a material and natural foundation – nor an industrial plant designed by engineers and organized according to a hierarchical principle in application of a plan-based model.

The third current adopts a socio-economic perspective, emphasizing the integration of the relationship to the natural world and its resources within institutions, cultures and moral views (Foster, 1997). This current is concerned with the development of institutions and social values that shape the formation of individual choices (see Kapp, 1950; Bromley, 1995), the way those choices are coordinated (see Godard, 1990), and the representation of the constraints imposed by nature on the human economy (see Godard, 2003). Far from being the passive outcome of individual preferences, human institutions are seen as shaping the emergence of those preferences by providing the conceptual points of reference, the knowledge and the information mobilized for their construction. Institutions also govern the distribution of profits, costs and risks (Beck, 1992), for example, by encouraging, through market transactions, the privatization of profits and the socialization of costs. This type of reversal of perspective – taking the institution or the system of legitimacy as a starting point for considering collective action and explaining the moral universe of individual choice (Godard, 2004) – was adopted in particular in the field of normative theories of development, with the birth in 1973 of the eco-development approach (see Sachs, 1980, 1998; Godard, 1998). Here, the central figures are not markets, contracts or individual preferences of consumers, but structures of decentralized, participative planning. Such structures make it possible for both the deliberation of citizen-producers and the engagement of their collective action to meet as closely as possible the specific needs and ecological conditions of each population.

Some authors, of course, may make contributions to more than one of these currents or put forward views combining elements of all three different points of view. Notwithstanding, the three currents described correspond to ideal theoretical stances that can be clearly distinguished.

THE FOUNDATIONS OF THE NEOCLASSICAL ECONOMIC REPRESENTATION OF THE WORLD

The neoclassical economic representation of the world[7] is based first on an elementary separation into two classes: individual human 'agents' – whose

preferences and choices supply the ultimate reference of judgement on allo-
cations – and 'goods' – useful objects, most of which are appropriated by
and exchangeable between agents, in other words, ordinary market goods.
Underlying this view of the world, there is the strictly anthropocentric idea
that a set of beings and objects – goods – are at the disposal of other beings
– humans, and humans alone – who have the right to use and dispose of
them as they see fit. Neoclassical economics is not the only school to adopt
this postulate, but it is totally permeated by it. The instrumental relationship it establishes belongs within the Western tradition that crystallizes
René Descartes' expression: 'to make ourselves . . . the lords and masters of
nature'. The complete development of the concept of property that emerged
in France from the French Revolution provided the practical quintessence
of this relationship by combining the right to use the good (*usus*), the right
to the fruits of the good (*fructus*), and the right to dispose of it (*abusus*).

The environment is then defined as a collection of goods or natural
assets that are useful to humans. Some, such as environmental amenities
(the spectacle of wilderness, landscapes, clean air, temperatures compatible with human life, and so on) are directly useful. Others are useful
through their incorporation into a production, in the form of factors of
production or raw materials (farmland, seeds, organic fertilizer, wood),
reserves of natural resources (forests) or functions of waste assimilation
(wetlands). Here, environmental goods share the same basic status as
reproducible goods (the reproduction of which is ensured by the apparatus
of economic production), although they possess some concrete particulari-
ties of which economic analysis endeavours to identify the impact on the
organization and regulation of the economy. As a result, this approach
treats man-made and environmental goods as commensurate, subject to
the same principles of evaluation and allocation and, more generally, as
substitutable elements.

In this representation, the human agents are the masters of the goods.
This entails the combination of two properties: first, the finality of the
goods is hetero-centred, meaning that they are exhausted in providing
satisfaction to human agents (mastery by destination); second, human
agents have legal and technical means to ensure their physical hold over
goods, to take them into their keeping (mastery by taking control). The
first of these two properties is expressed most obviously when the goods
are designed and produced with a view to satisfying the needs and desires
of agents. When applied to self-organized and self-finalized natural beings,
on the contrary – especially living beings – it requires efforts of domestica-
tion to gain control over them and divert them from their own ends. The
second property entails the mobilization of knowledge, technical know-
how and instrumental power, which is used to take control, for example,

by diverting a river, taming wild animals, or transforming animal life into an industrial production of food and waste.

Whatever one's opinion of the views it may accommodate, the environmental ethic has the immense merit of presenting other possible representations based on other distinctions than those of the neoclassical representation. One of them, in the tradition of Jeremy Bentham, extends the concept of utility to non-human, sentient beings. Others affirm that these beings have rights that we should acknowledge, at least in the form of human obligations towards the non-humans. Others, through ecocentrism, argue for the integrity of the living world, of which the human race forms a part, as the superior good that we should aim for.[8] Depending on the case, these reflections concern either more advanced animal species, all species, with in that case the anti-specist postulate that all species have the right to experience their natural life potential in accordance with their biological constitution, or biotic communities associating living beings with the physical substratum (ecosystems, biomes, or the biosphere as a whole). If we take a step in any one of these directions, it is no longer possible to reduce non-human beings to the status of goods to be exploited or destroyed for the sole benefit of humans.

Here once again, it is clear that the issue underlying the environmental question, even in a philosophical context free of any religious reference to divine creation, is the legitimacy and effectiveness of the mastery of human agents over the rest of the world. Before becoming an issue of management, the question of the environment was brought to the fore by a movement criticizing the technical, economic and social rationale that has governed the development of Western capitalism since the beginnings of industrialization. This criticism, deriving from various origins, is levelled at the two components of mastery: the enslavement of natural beings to the end purpose of human consumption, and the effectiveness of capacities of control, both over nature and over the technical objects that humans introduce into the environment (for example, chemical pollution with synthetic substances, diffusion of genetically modified living organisms).

In the first case, the issue is the pure and simple reduction of nature to the status of a collection of 'goods'. In the second case, the realization that humans are incapable of controlling the effects of their technical creations calls into question the practical effectiveness of their mastery: there are always dispersion, leaks, explosions, contaminations, and therefore environmental degradation. Thus, the environmental crisis stimulates reflection on the exercise of human mastery over the world. What would reflexive mastery be like – a mastery that took into account, in its relation to the world, the knowledge of its own limits in terms of both end purposes and controlling the effects of human actions?

Against the backdrop of these questions, it is no longer possible to see the application of the neoclassical economic representation to environmental issues as just another triumph for rationality. Does it not start with a powerful reiteration of the subjugation of the world to individual human subjects? Does it not lay itself open, consequently, to the risk of remaining blind to the source of the profound unease provoked by the fact that human development – as it has been known since the nineteenth century – has severely put the biosphere and the global environment in peril? Is it not an arrogant expression of the refusal to reconsider the idea of human mastery? From a structural, long-term viewpoint, is it not more a part of the problem than the solution?

To explore these questions further, I propose to describe in more detail what is at stake in the environmental relation, and then to consider the economic representation of the world as an operator producing effects on the natural environment. I shall dwell particularly on the consequences of the dominant, neoclassical economic representation.

THE CONCEPT OF ENVIRONMENT DERIVING FROM THEORIES OF THE SELF-ORGANIZATION OF COMPLEX SYSTEMS

What do we mean when we describe a good or a problem as being 'environmental'? I propose to explore this question with reference to works on the self-organization of complex systems (see Godard, 1997a). This conceptual framework will enable me to situate clearly the theoretical position occupied by the different economic currents identified above.

The Environment and the System: An 'Entangled Hierarchy'

Let us start by looking at the cognitive and practical relations established between a self-organizing system and its environment. To this end, we postulate the existence of a system whose identity and autonomy are the joint result of the self-maintaining activity of an operationally closed network of components[9] and a physical opening onto a domain of existence that constitutes its environment. The idea of operational closure means that this type of system creates its own space of meanings and gives it an organizational base. The self-centred rationality resulting from this is the principle through which events, fluctuations and changes in its environment take on meaning for it as disruptions or opportunities. By situating this system in its environment we can connect its existence, its activity and its future to an encompassing interactive context that both includes and surpasses it. The 'environment' pole thus brings together a set of critical conditions for the existence and development of the system.

The operations of identification and delimitation of the system's environment, which demarcate at the same time the system's contours and domain of existence, proceed logically from the nature of this system chosen as the object of reference and from the relations that it establishes with the surrounding world. The category of environment – whether we are dealing with Nature or ecosystems – can only designate those things that take on a meaning for a system of reference. From this perspective, the environment appears as an extension of the system, its complement, or counterpart. At the same time, the category of environment connotes an exteriority in relation to the system of reference. This concept of exteriority is intended to account both for the limit of the domain of practical and informational control acquired by the system and for the difference between its domain of action and its domain of self-reproduction, the latter being more restricted. Three consequences can be drawn from these premises.

1. The environment of the system represents a critical context for the reproduction, survival and development of the system; on the contrary, the objects and beings outside the system only truly constitute the environment of the system if they represent, alone or together, elements that are relevant to the system.

2. To the extent that the autonomy of the system is based on the definition of its domain of self-reproduction,[10] what we call the environment is then constituted of objects and forms of organization whose reproduction is not actively and regularly incorporated into the organization of the system itself. The active movement of self-reproduction of the system (maintaining of an invariant structure; permanent renewal of physical components) is at the same time the active mechanism by which the 'externalization of the environment' is generated, that is, the process by which – because of its organization – the system rejects the systematic responsibility for reproducing a given component or resource. The autonomy of the system and the externalization of its environment are thus two sides of the same coin.

3. Considered at the meta-levels where coherent logics of organization can be observed, the components of the system's environment are found in their interactions with each other and with other elements of encompassing systems from which emerge specific autonomous logics that create their own meaning (dynamic equilibria, regulations, fluctuations). Therefore meta-level systems represent a permanent source of uncertainty and surprise for the reference system.

Through this last aspect, a hetero-referential element is introduced into what starts off as a self-referential logic emanating from the reference system: the latter cognitively and practically 'constructs' its environment through what it is itself, but this environment also represents a source of fluctuation and threats to its activity and survival, which it does not control and to which it must adapt. Some of those threats and constraints result through a feedback loop from the very process of externalization that it imposes. The reproduction of the environment cannot be taken for granted. In fact, the primary tendency is the one through which the system gradually undermines the conditions of reproduction of its environment. When self-reproduction is finally achieved, it appears as the risky result of an externalization being compensated and overcome by the means of adaptation and evolution, well illustrated by the phenomenon of co-evolution. If we consider the overall regulation of the relationship of the 'system–environment' pair, we can record the existence of a gap, bearing a constant threat to the survival of the system. It is also this gap, however, that sets the relations between the system and its environment within an open history that can generate novelty, rather than a never-ending, machine-type repetition.

Under these conditions, the central theoretical proposition put forward is that the system–environment pair organizes two opposing hierarchies of meaning[11] H and $\underline{H}$, which determine the constitution of the entities involved (the system S and its environment E). The pair therefore has a structure of 'entangled hierarchy'.[12] By defining the relation > as a relation of logical and/or definitional precedence, I aim at the following questions. What proceeds from what? Which entity is the condition of the other? We can then characterize the structure of the entangled hierarchy (S–E) as follows.

1. The pair S–E paradoxically associates two opposing hierarchies of meaning H and $\underline{H}$, with H such that $S > E$, and $\underline{H}$ such that $E > S$ (1).
2. This association is such that:

    - from a structural point of view, each hierarchy includes the other as a necessary component of itself (2);
    - from a dynamic point of view, each hierarchy logically and provisionally leads to its opposite, constituting a moment of it, in a circular movement analogous to the circumnavigation of a Moebius strip (Figure 2.1) (3);
    - ultimately, in the case of relations between the human economy and its natural environment, the entangled hierarchy is dominated cognitively by the hierarchy H: $S > E$, because the ques-

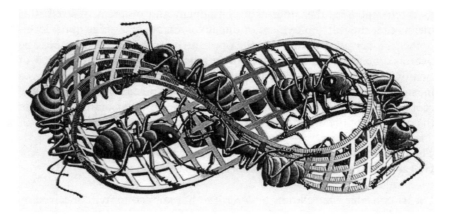

Figure 2.1   Escher's Moebius strip

tion of the environment can only be raised by humans, but it is dominated practically by **H**: $S < E$, in so far as the existence of nature is a logical precondition to the existence of humanity.

Proposition (1) follows from the following two points.

- On the one hand, the first term in the hierarchy is the system S initially chosen as primary reference in the analysis; the environment is the second term, because what constitutes it as environment depends on its relation to the system of reference. This hierarchy H corresponds to the affirmation of the *self-referential* logic emanating from the system. Here, the point of view of the analysis seeks to coincide with the point of view of the system itself.
- On the other hand, the environment, as an encompassing reality from which emanate coherences, regulations and essential resources, but also threats, disruptions and disequilibria, imposes itself on the system as a precondition to its existence and survival. With this status, the opposite hierarchy H is affirmed, where the environment is the first term and the system is the second, dependent term. H corresponds to the *hetero-referential* logic in which the system is subjected to a reality producing a meaning that is not its own, but in which it participates and on which it is crucially dependent.

Thus, the paradox of the conceptual pair $S–E$ lies in their association of two hierarchies that oppose each other without cancelling each other out. Propositions (2) and (3) concern the effects of the paradox: each hierarchy includes the other as a component and moment of itself; this means that the full deployment of one involves, at some stage in its development, a temporary shift into the opposite hierarchy. On the whole, the overall dynamics emanating from this pair does not lead to a space of homogeneous meaning, but to a complex meaning: it organizes spaces of meaning built on opposite relations. This can be precisely observed in the economic research programmes that oppose each other diametrically on the question of the biophysical environment of human societies.

The Distribution of the Three Economic Currents in the Entangled Hierarchy of the 'System–Environment' Pair

Let us consider the relation between the human economy, conceived as a complex system capable of self-organization, and the 'biosphere', conceived as the environment of that economy. The self-referential point of view finds pure, reductionist expression – it is the core of methodological individualism – in the neoclassical representation, as the search for the optimal allocation of goods is ultimately subject to the preferences of the contemporary individual agents who make up the society. Furthermore, the theory of externalities,[13] which serves as its main framework for tackling problems of pollution and ecological disequilibrium, considers transformations of the environment as nothing more than vectors of an uncompensated variation in agents' utility, ensuing from the existence of 'off-market' interactions between agents.

If we apply this conceptual structure to the environment, with a view to 'internalizing the externalities', and we follow the logical steps, we are led to the inescapable conclusion that this type of externality can only be recognized, evaluated and taken into account if humans first try to identify with the environment – that is, if they adopt the hetero-centred point of view – in order to determine the ultimate systemic meaning of this process for the environment itself. Then, and only then, could it be assessed from a human perspective. In other words, to qualify an external damage in economic terms, we must start by revealing the sequences of causality between the biophysical phenomena set in motion, so as to identify the final effect incurred by the environment. Only then can we qualify in return the way the utility functions of the human agents have been or will be affected. Here, experts and agents are obliged to abandon the self-referential semantics of economics (preferences, utility, choice, evaluation, price and so on) in favour of a hetero-referential semantics (energy transfers, dynamics or disequilibria of species populations, functional disruptions, instabilities, resilience and so on). At that moment, the environmental meaning emanating from the biophysical processes in play is in command, controlling the economic meaning that ultimately depends on it. The self-referential logic must shift temporarily into hetero-reference as a moment of its own realization. To do his job, the economic expert must therefore translate from one semantic universe to the other twice: first, from the self-referential meaning to the hetero-referential meaning, and second, vice versa. These translations are tricky, because the two semantics are no easier to match than the frameworks of data collection.

'Bioeconomics' or 'ecological economics' adopts a position that is symmetrically opposite to the neoclassical approach. Through identifying with the hetero-referential point of view,[14] it seeks to express the natural environment's reproduction constraints that should be acknowledged by the human economy to make it sustainable. In fact, the researchers who adopt this point of view are not in a position to know directly and perfectly the meaning that biophysical phenomena have for the environment. They approach that meaning from the construction of scientific knowledge that can only be partial and is marked by the historical and social contingency specific to the human society to which they belong: dated state of development of knowledge and techniques, distribution of resources among research institutions, public policy priorities for research, forms of organization and circulation of knowledge – including the issue of the extension of property rights or, on the contrary, of 'open access' style of public diffusion – the influence of ideological currents and trends, and so on. For that reason, scientific statements are conjectural and historical emanations *of the human economy* (here I consider the production of knowledge as an integral part of the human economy) in the very movement whereby they claim to deliver the meaning of the phenomena in question *for the environment*. Consequently, no interpretation from a hetero-referential viewpoint can escape from the process of scientific and social controversies (see Godard, 1997b) or from the incompleteness of statements about the meaning of the environment as a whole. The hetero-referential interpretative frameworks are indelibly marked by the self-referential point of view.

Like other types of human reality, the representation of the environment is constructed socially through a compound of social representations, forms of interests and diverse types of knowledge of which scientific knowledge is but one component. This representation depends on frames of interpretation and particular institutions. The point of departure chosen by environmental socio-economics is to challenge the two symmetrically opposed reductionisms – the individualist self-referential on the one hand, and the holistic hetero-referential on the other – to show how the two hierarchies interact and 'lean on' each other, echoing the way Cornelius Castoriadis (1975) defined the relation between the human institution and natural processes.

Taking into consideration humans' uncertainty with regard to the environment, environmental socio-economics explores the collective processes by which environmental problems are defined and dealt with at the interface between public debate, state action and economic processes. The issues of institutional change, the legitimacy and justification of the public action, coordination and social ties and the conditions of commitment are emphasized, but without disconnecting them from the classic economic issues (formation and distribution of income, competitive strategies, allocative efficiency and so on). This socio-economic approach repositions economics in its socio-institutional context and at the same time within the biophysical materiality of the world. Equally, however, it seeks to account for the opposite movement, by which economic mechanisms select and give life to certain issues of legitimacy and institutions, but not others, and produce the physical world of tomorrow under conditions that make the reproduction of the environment an open question through which humanity endangers itself.

In conclusion, if the two opposing poles (neoclassical environmental economics and bioeconomics) can each be clearly and self-proclaimedly identified with one of the constituent hierarchies of the entangled hierarchy, environmental socio-economics operates at the junction between the two. Through the concept of 'leaning on', it seeks to explain the relative autonomy of the human economy in conceiving problems by highlighting the plurality of possible and actual constructions. It also seeks, however, to reveal the importance of the foundation provided by the natural phenomena themselves. It is therefore not surprising that this socio-economic current should be so interested in situations of scientific controversy and processes of scientific expertise under uncertainty.

Finding Unity in the Combination of the Three Currents of Environmental Economics

To understand the environmental field as a whole, we must mobilize knowledge produced by all three of the currents identified above, even if their constituent hypotheses are different, or even contradictory. We are faced with a remarkable situation in the way that this object – the environmental field – imposes itself on the approaches to learning it. Here, to be understood, the object does not impose the formation of an integrated, semantically coherent theoretical field, but three conflicting components, which must be taken together to render the object fully intelligible. It is within the cognitive matrix formed by the system–environment pair that the fundamental features of the object of our inquiry and the questions it raises can be revealed. Here, we are far removed from the empiricism by which the concrete object imposes itself on knowledge without depth or mediation. We are also far removed from the constructivism according to which the object is created entirely *ex nihilo* by the premises of an approach to knowing it, to the point where the same object can be the subject of an infinite number of different constructions.

The conception proposed for gaining understanding of the field of environmental economics indicates that adequate knowledge of the essential issues attached to this object requires us to circulate between the different points of ontological identification, therefore entailing semantic breaks or shifts. It shows us a mode of integrating different elements of knowledge by circulating between points of view that remain distinct. In addition, this type of integration cannot be reduced to a process of convergence of previously acquired, self-founded and self-sufficient knowledge. Here, the circulation between points of view is necessary if each approach to knowledge, based on one of these conflicting points of view, is actually to succeed.

THE NEOCLASSICAL THEORY OF INTERNALIZATION OF EXTERNALITIES AT ODDS WITH SUSTAINABLE DEVELOPMENT

Since the early1920s, the Pigouvian tradition of public economics expects the internalization of negative externalities to be achieved through corrective action by the state. Consequently, a tax reflecting marginal external damage – which therefore needs to be evaluated – is the favoured theoretical instrument for aiming to achieve optimal internalization. Since the1960s, the Coasian tradition has reinterpreted the problem in terms of incomplete property rights and, once these rights have been defined, relies on the contract and the market to spontaneously reinternalize the externalities of which the social cost is higher than the transaction costs imposed by internalization. The role of the state is not done away with, but it is different, because it consists essentially in creating and protecting property rights.

In both cases, beyond the choice of instrument, the problem raised is that of determining the level at which 'internalization' should be taken. For example, what level of taxation on CO_2 emissions should be introduced into the economy? Alternatively, at what level should we fix a cap on GHG emissions, which is then spread out in the form of emission quotas for companies? The self-referential, neoclassical solution is to choose an 'optimal' level that equalizes the marginal external damage incurred by individual economic agents and the marginal cost of reducing emissions.

This solution seems to make good economic sense; it is, however, an illusory solution. Admittedly, when compared to a situation where these externalities are not taken into account at all, it will mitigate the process of environmental deterioration, but it will not stem the process. In other words, the neoclassical way of conceiving the internalization of externalities is part of the general process by which the human economy externalizes its environment and subjects it to a cumulative process of deterioration. This theory thus cannot claim to provide the foundations for policies of sustainable development.[15]

Dynamic Externality, according to David Pearce (1976)

A model proposed by David Pearce (1976) more than 30 years ago will serve as our point of departure for this reflection. The essential data of the argument are presented in Figure 2.2.

Let there be the production of an economic system resulting in a quantity X of a product and pollutant waste R, such that the quantity of waste is an increasing function of production ($R = R(X)$ and $R'(X) > 0$). This waste is dispersed in the environment, which we assume to have a carrying capacity A_t, defined as the quantity of waste R that it can absorb, break down and return into natural cycles without damage during the period t (for example one year) when the capacity A_t is available at the beginning of the period t. The function $R(X)$ and the capacity A are represented in the upper part of Figure 2.2. The lower part of the figure represents the function B^m of the marginal private profit obtained from the production, excluding external costs. This function depends on the level of production X. This part of the figure also represents the function of marginal external damage D_t^m resulting from the emission of waste during the period t, taking into account the carrying capacity A_t available at the beginning of the period. This damage function is increasing with the level of waste and therefore, in this model, with the level of production.

At the beginning of period 1, the available carrying capacity is A_1. The 'private' optimum production is X_p, when the marginal private profit obtained from production becomes nil. The pollution emitted causes the external damage represented by the curve $D_1^m(X)$. According to the neoclassical method of determining the optimal internalization of these external costs, we must compare the damage suffered by the 'victim' agents with the profit represented by the curve $B^m(X)$, and which must be renounced to limit the level of waste and, therefore, of external damage. The result is an optimal level of pollution R_1^* and a corresponding optimal production X_1^*. Yet what happens to this pollution, which is deemed to be optimal but which exceeds the environment's carrying capacity A_1? Because of the laws of the conservation

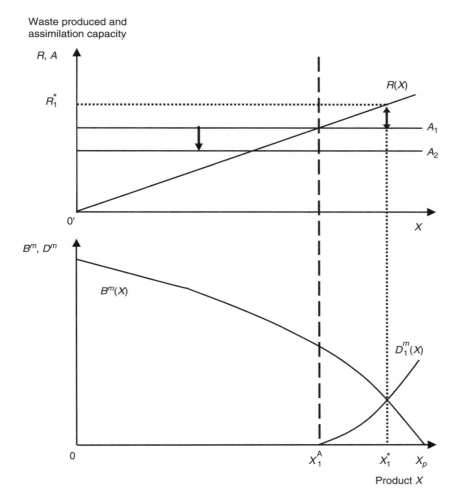

Source: Pearce (1976).

Figure 2.2   The problem of dynamic externality: the first period

of matter and energy and of ecology, this excessive pollution reduces the carrying capacity in the following period. This is shown in the upper part of Figure 2.2 with the lowering of A_1 to A_2, the new carrying capacity available for period 2. This process is recurrent. The carrying capacity falls a bit further in each period 2, 3, . . ., n, and the marginal external damage curve shifts towards the left, which in turn shifts the optimum production towards

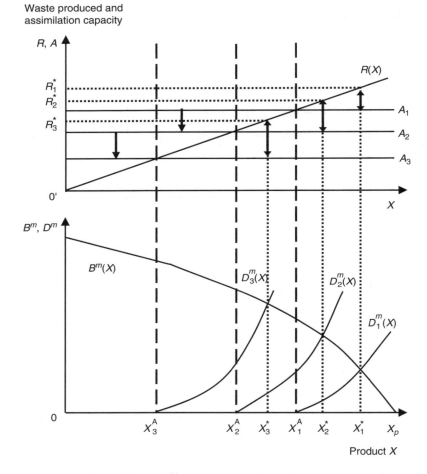

*Figure 2.3   The problem of dynamic externality: the process over three
periods*

the origin, from X_1^* to X_2^*, then to X_3^*, . . ., X_n^*, and the optimal level of waste
from R_1^* to R_2^*, then to R_3^*, . . ., R_n^*, as illustrated in Figure 2.3.

Intuitively, the long-term outcome of this process is clear: after n periods,
the whole assimilation capacity of the environment will have disappeared,
and with it, any possibility of production. If we feel that there is some veri-
similitude to this chain of events, the neoclassical method of internalizing
externalities thus leads to the ultimate ecological catastrophe. But perhaps
Pearce's model is false?[16]

To explore this question, we should start by grasping the logic of the model, which owes nothing to the contingency of the graphic representation. The decisive point is this: in the lower part of Figures 2.2 and 2.3, the external damage curve $D^m(X)$ originates at point X^A, the maximum production compatible with the carrying capacity available for the period under consideration. The reason for this is that below this threshold, the environment assimilates the waste emitted without suffering any functional deterioration. There is therefore no resulting damage for the 'victim' agents; the waste does not generate any externalities in the sense of the theory under consideration. It is only when the threshold capacity A has been crossed that the damages start to be felt by the different economic agents: loss of natural resources, health incidents, the extra economic cost of satisfying essential needs like supplying the population with water, and so on. By theoretical necessity, the optimum production after neoclassical internalization of environmental externalities therefore exceeds the threshold compatible with the carrying capacity of the environment.

Different solutions have been envisaged to avoid this fatal course of events, including technical progress, potentials of adaptation, or so-called 'no-regret' or 'double dividend' strategies, which make environmental policy an accidental windfall from the pursuit of other objectives (see Godard, 2006), but none of them is really convincing. Here I shall only consider the case in which the environmental damage does not result from an annual flow of pollutant emissions exceeding a steadily decreasing carrying capacity, but from the accumulation of pollutants in the environment. The problem of climate change corresponds to this case: a proportion of the gases emitted accumulate in the atmosphere, with the rest being recycled by the oceans and vegetation. The accumulated stock modifies the thermal equilibria between the solar radiation received by the planet and the radiation emitted by the Earth's surface. Here, the environmental damages depend not only on the scale of the changes involved, but also on the speed at which they occur, which may be too fast for the capacities of adaptation of species and ecosystems, and for human populations.

As the annual addition of pollutants exacerbates the total damage, the marginal damage associated with any given level of addition to the stock of pollutants accumulated in excess of the assimilation capacity of the planet becomes higher from one period to the next. Not only does the curve $D^m(X)$ shift towards the left, but it also becomes steeper. Within the context of intertemporal optimization by a benevolent and omniscient planner, this steepening can partly regulate the tendency to emit pollution in each period, because the arbitrage between marginal damage and marginal cost of abatement also shifts. In stock pollutant models, where nature itself is assumed to regenerate its carrying capacity at its own rhythm, in the form of elimination of part of the stock of pollutants, some models[17] suggest that it is possible to end up in an equilibrium where the combined action of natural purification and rising damages results in pollution being held at a level where the residual carrying capacity is maintained. Nevertheless, there is nothing systematic about the emergence of this equilibrium, and even if it does appear, what level of environmental deterioration would we have to accept to get there? David Elliott and George Yarrow (1977), for example, presented, as a response to Pearce, a dynamic pollution model bringing to light the conditions of equilibrium of a pollution stock, in which they hypothesized that the equilibrium between incoming pollution and outgoing pollution (stock model with inputs and outputs) could maintain the environment and its carrying capacity in a stationary state. These conditions were determined by the simple equation:

$$\frac{\textit{Marginal cost}}{\textit{of abatement}} = \frac{\textit{marginal damage of the stock of pollutants}}{\textit{discount rate} + \textit{marginal rate of natural elimination}}.$$

Can we be satisfied with this type of equilibrium?

The Proof by Climate Change

Elliott and Yarrow's equation highlights the crucial role played by the discount rate, which mitigates the presently perceived severity of future damage. This aspect has taken on critical importance in the economic debate provoked by the publication of the Stern Review (Stern, 2006) on the economics of climate change (see Godard, 2007a, 2008). The choices made by Stern[18] to evaluate the damages up until the year 2200, compared to those made by William Nordhaus in his own modelling,[19] lead to results that differ by a factor of 10 for the present value of a tonne of CO_2 to be taken into account to determine climate policy. With a discount rate reflecting the current functioning of the capital markets – in application of the neoclassical postulate that the market reveals individual preferences, that is, a rate of about 6 per cent net of inflation, according to Martin Weitzman (2007) – the solution proposed is obviously only formal and absolutely not realistic in the face of the risks of climate change.

The processes of climate change are such that even if there were international agreement on a world target of zero GHG emissions from 2010, it would take several tens of thousands of years to restore the climate conditions of the mid-twentieth century. Moreover, Weitzman (2009), drawing on the most recent climate modelling and data from the 2007 IPCC report, estimated that a scenario at 550 ppm (parts per million) of

CO_2-e (carbon dioxide equivalent) atmospheric concentration of GHG, which is the limit not to be exceeded according to the Stern Review but is often considered to be an ambitious target, nevertheless translates into a 5 per cent probability of seeing a global mean temperature increase of 11°C or more in the long run. This is a considerable level of probability. At present, no one can say what our planet would resemble under such conditions. The main forms of life, including humans, could disappear. Yet on the basis of recent trends in the rising atmospheric concentration of GHG, we shall reach this value of 550 ppm somewhere between 2035 and 2050.

Climatologists largely agree that the level of GHG emissions that can be offset by an equivalent absorption by the natural carbon sinks of the oceans and vegetation is about 8 to 10 Gt (gigatonnes) of CO_2-e per year (Le Treut, 2004, p. 20), that is, between one-fifth and one-sixth of world emissions in 2005. Using cost–benefit analysis, the Stern Review, nevertheless criticized for its pessimism or radicalism, recommended a target concentration of between 450 and 550 ppm, and in fact closer to 550 ppm than 450 ppm. This would mean aiming for 20 Gt CO_2-e emissions per year in 2100 following a trajectory of gradual worldwide absolute reductions starting in the 2020–30 decade. This scenario would nevertheless be accompanied by a probable mean temperature increase – with all the resulting climatic imbalances – of about 3°C compared with pre-industrial levels. It would still be far from a level that would guarantee the climatic sustainability of economic activity. The climatologist James Hansen, director of the NASA Goddard Institute for Space Studies, recently declared that anything more than a temporary overshoot of 350 ppm CO_2-e, which we have already been above for the last 20 years, would lead humanity into a dangerous climatic adventure (Hansen et al., 2008), in contradiction with the objective agreed by the international community in Rio de Janeiro in 1992 when the Framework Convention on Climate Change was adopted and brought to the forefront of the Earth Summit: 'To avoid dangerous human interference with the climate system'.

WHAT SOLUTIONS?

The Uncertain Hope from the Heuristics of Fear and Green Technologies

In a world combining a fragile environment with limited carrying capacities and a human economy lacking the regular and effective exogenous technical progress necessary to ensure constant improvement of its total environmental performance,[20] human production will only be contained within the limits of environmental assimilation by powerful integration of the threat of a catastrophic climate change into the collective utility function. This certainly cannot be achieved by limiting the horizons of public policy to the current functioning of financial markets! On this point, such a requirement echoes the philosophical construction of Hans Jonas (1984) around a heuristics of fear to make current generations aware of the apocalyptic threat that human action poses to the future of the world and the future of humanity. Jonas, however, despaired of the capacity of democratic societies to attach enough importance to future risks, because the future has no lobby.

An alternative to shocking people into awareness of the threat of catastrophe would lie in an exogenous technological progress intrinsically beneficial to the environment and occurring regularly and in abundance. To fulfil this role, technological progress would have to take very specific forms over the long term to avoid the fundamental argument of Jonas and other critics of technology, according to whom technology is the main source of the apocalyptic threat, because it radically increases the human power of action over the world and becomes embodied in systems that escape collective human control. Developing a profile of technological progress that would dramatically reduce the human impact on the functioning of natural systems would thus be the great challenge for the twenty-first century. This cannot result from the neoclassical approach of internalizing externalities, and cannot be envisaged without the establishment of a powerful system of social regulation of the interface between science and technology and of the conditions of implementation of technological innovations.

For this approach of controlling technological development to be possible, however, the authorities must be capable of distinguishing at an early stage between technological developments that are likely to present a grave threat to the environment and those that are inoffensive or beneficial. Without this capability, there can be no guarantee as to the solution provided by technological progress. It is, however, precisely the impossibility of making this distinction at an early stage that undermines the validity of the maxim proposed by Jonas for apocalyptic risks:

> Never must the existence or essence of man as a whole be made a stake in the hazards of an action. It follows automatically that here, simple possibilities of the type that have been described [those with apocalyptic potential] must be considered unacceptable risks, which are not rendered any more acceptable by any of the opposable possibilities . . . We must treat that which can certainly be put into doubt, as long as it is possible – once we're dealing with a possibility of a certain kind – as a certainty with respect to our decisions. (Translated from Jonas, French edition, 1990, p. 62)

Without the necessary knowledge, Jonas's injunction cannot avoid arbitrariness in the choice of targets or the paranoia of widespread 'catastrophism' (Godard et al., 2002; Godard, 2007b). *Ex ante* we must resolve to accept technological progress in its fundamental ambivalence, which limits, without completely destroying, the hopes that we can invest in it.

This can only happen in an environment capable of self-regeneration, despite the pressure from humanity that the standard theory of externalities recovers its relevance by only dealing with 'human affairs'. This was in fact the hypothesis of the classical economists who laid the foundations of the theory of externalities.

A Theory in Need of Revision

Economists must carry out a profound revision of the neoclassical theory of externalities[21] if they want to incorporate it into the perspective of sustainable development. If we use the term 'externalization' to designate the movement by which a system alters the conditions of reproduction of its environment, then we can say that the standard mode of internalization proposed by this theory incorporates externalization into the very heart of the proposed internalization. In that respect, this theory simply embodies the self-referential logic of a human economy lacking all reflexivity about its insertion in the natural world. As is true for any system that, having defined its own boundaries, externalizes (that is, does not take responsibility for) the issues concerning the reproduction of its environment, the destiny of this economy will be, for better or for worse, to have to adapt to the environment that will have been spoiled, exhausted or at the very least profoundly transformed.

To invent another future, we must return to the source of the error in handling a conceptual scheme that was originally constructed for another purpose. The error lies in the way that the neoclassical scheme of internalization compares two types of costs that are treated symmetrically: external environmental damages and the economic costs of reducing pollution, which are incurred by economic agents and integrated into the formation of market prices. But these two types of costs actually possess a conceptual asymmetry that makes it impossible to consider them as being directly commensurable. This asymmetry stems from the fact that internal economic costs are borne because they will be compensated – in a movement of reproduction of the economic conditions of production by the creation of a market value – and the realization of this value in the exchange allows for the renewal of the productive cycle. At economic equilibrium, eliminating the risk factor for companies, each agent finds the conditions at the end of each period that enable him or her to start the cycle over again in the next period.

Conversely, what is designated as 'external damage'[22] corresponds to a net destruction, consisting in a fracture of the regulation and loops of reproduction of the biophysical systems affected. The neoclassical concept of externalities takes into account only the secondary consequences of this uncompensated net destruction on the utility functions of agents. Written in from the start as a postulate of the argument, the net destruction of the environment can be found again at the end of the argument, but without ever having been validated. The use of this theory to define environmental policy amounts to accepting, knowingly or not, a meta-decision establishing a division between the factors and goods for which the modeller integrates a constraint of reproduction (the classic factors of production: capital and labour) and the factors and goods for which he or she rejects such a constraint.

In the case of productive capital, the concept of reserve for depreciation was invented to ensure that the economic process renews the economic resources required to replace or reconstitute worn or damaged physical capital. The question then arises: what is the justification for the economic approach to the environment of not imposing a similar concept on natural capital? Here, a primary decision must be made to set the economic development of humanity within an environment whose constraints of reproduction are taken into account structurally, in the same way that the constraints of reproduction of 'classical' economic factors are taken into account in microeconomic, as well as macroeconomic, models. Why adopt a position with regard to the environment that would appear clearly absurd if we adopted it for productive capital?

Admittedly, it is possible to envisage a meta-decision of externalization or ecological destruction being taken on a local scale, but such a decision must be well enough informed to ensure its validity. The neoclassical theory of externalities cannot be used to inform the decision, because its premises place it downstream, when the choice has already been made not to consider conditions of environmental reproduction. On a global scale, on the contrary, such a meta-decision can only lead the human experiment to an apocalyptic outcome.

Rehabilitating the Concept of Reproduction

The guiding idea that I propose is to rehabilitate the concept of reproduction, over and above concepts of equilibria, and to extend this concept to cover the environment, while bearing in mind that extension of the domain of self-reproduction of the human economy can never mean more than pushing back the boundaries: it can never abolish the founding 'system–environment' duality. For this purpose, new constraints of ecological reproduction need to be established and imposed on economic thinking, evaluation and activity. Only values that are compatible with the reproduction requirements of the new extended domain should be accepted for the purposes of equivalence and compensation. The foundations of the comparability with the economic costs of reducing pollution would thus be restored. Economists would then be justified in using their methods to determine, for development projects or public policy choices, which of the two strategies is more efficient: reducing pollution at the source, or letting it follow its course and then meeting the costs of cleaning up and restoring the environment. In concrete terms, this means that in Figures 2.2 and 2.3 the function $D^m(X)$ should be replaced by another: $E^m(X)$, representing the marginal cost of restoring the environment in its functions and regulations. In other words, we must invent the equivalent of the concept of reserve for depreciation to ensure the reproduction of the environment.

True, this concept of reproduction is difficult to use when we seek to give it an empirical content: ecosystems evolve; the question of the thresholds of disruption below which they can reproduce and above which they deteriorate is controversial and depends on the nature of the environmental issues at stake; individually exhaustible resources cannot be reproduced within a relevant human horizon, but technological potentials of substitution may exist for some of them, for example. Here we come back to what justifies the pivotal role of a socio-economic approach working on the pragmatic interpretation of biophysical processes and their insertion within the strategic games of economic and political actors and within the orders of justification at work in human societies (Boltanski and Thévenot, 1991; Godard, 2004), but also on the standards of intra- and inter-generational fairness to be promoted in the context of policies of sustainable development.

This pivotal position is strengthened, moreover, in situations far removed, because of past behaviour, from environmental sustainability. This is obviously the case today with the prospect of global climatic upheaval. Scientific and technological knowledge gives access only to fragmented, indirect and partial approaches to the functional limits and viable thresholds of the environment. Despite the best efforts of organizations like the IPCC to establish a shared assessment of the knowledge available, science does not provide intangible and imperative benchmarks that compel unanimous acceptance. Consequently, not only do human communities have to seek to discover nature's limits, but they must also decide on the limits they set on themselves in organizing their development, reflecting the possible but uncertain limits that they attribute to nature. And these decisions require detailed discussion, because the outcome affects everybody, whether on the micro-regional scale of the management of a water table or on the global scale for the climate. As a result, assessment of the limits depends not only on the knowledge available about nature and about the technologies that can be used, but also on collective preferences. It cannot avoid taking on a normative dimension, where assessment of the technological and economic possibilities nevertheless has a role to play: one of the constants of decisions under scientific uncertainty and controversy is that the solutions adopted are shaped more according to operational technological potentials within a horizon relevant to the decision makers than according to an objective and scientific definition of the problems to be resolved (Godard, 1997b).

CONCLUSION: MANAGING THE UNSUSTAINABLE

Since the international community has not yet proved by its actions that it considers climate risk to represent a serious threat of global catastrophe that could endanger the survival of humanity, the radical options that might have averted climatic upheaval 20 years ago were not taken. The current challenge is not to maintain economic development within the limits of environmental sustainability. It is to determine by what means and at what speed we can embark on a transition towards a state that would be less unsustainable over the long term, without any guarantee, in doing so, that we would avoid a catastrophic change in the environment, given the irreversibility of the deviations from sustainability that have already been made.[23] I hope that the reader will forgive me for using a biblical image: the original sin has been committed and humanity has been driven out of the paradise of environmental sustainability. This took place, moreover, at a time when Europe was looking back nostalgically on the 30 years of post-war economic boom and the developing or least-developed countries were wondering how to catch up with them one day. No optimizing algorithm can give us a solution to the problem of finding a way out of unsustainable development.

In any event, we know that we must reject two models that claim to deduce the collective decision to be taken from objective knowledge of 'the nature of things', the first of an ecological nature, the second economic.

- The first is a model in which resources are optimized within the context of limits laid down objectively based solely on declarations emanating from the natural sciences. Due to the controversies and uncertainties that remain unresolved, the scientific representation of the climate problem cannot provide a unique, firm benchmark on which to build a global strategy: what target should we aim for, 350, 450, 550 or 750 ppm, and following what trajectory?

- The second is a model in which the best policy is identified through economic calculation based on a table of costs and benefits, following the standard schema of neoclassical theory, because this framework is fundamentally biased when deterioration of the essential functions of the environment is at stake.

We also know that to understand the situation one must circulate between and test by trial and error the points of view that have been referred to in this chapter as 'self-referential', 'hetero-referential' and 'pivotal', taking care to identify the semantic discontinuities and shifts that this exercise requires.

In practical terms, and in relation to the contribution of economists to public choices, this brings us closer to the position set forward by William Baumol and Wallace Oates (1975) 30 years ago concerning environmental policies. According to these authors, the definition of the objectives to pursue in the field of the environment is a scientific and political process lying outside the competence of economists. Nevertheless, the learning of economists should be called upon to enlighten the choices of the best instruments for attaining these objectives. Contrary to appearances, cost-efficiency analysis is better adopted than cost–benefit analysis here, because it makes it possible to rightly account for the ecological and social realities in their own terms. Cost-efficiency analysis is a more cautious instrument that expresses doubt about the pretention of conventional welfare economics to determine an all-embracing, optimal, long-term future of humanity as if it could be the sole result of an objective account of contemporary individual's preferences revealed on the market place or through market-like methods.

This more modest methodological position must nevertheless be enriched by giving a central role to one of the key innovations of the concept of sustainable development, which is to ask present generations to make sure that they protect the *capacities* of future generations, echoing one of the major themes in the work of Amartya Sen (1999). One operational application of this precept would be, given the temporal limitations of markets and uncertainties about the nature of the future world, to abandon the idea of optimizing long-term development trajectories in favour of a sequential approach covering successive periods of about 30 years (one generation), more easily controlled by the economic and political decision makers. Attentive to the revelation of information and the need to adapt choices according to that learning process, this approach would take particular care over the end-of-period conditions transmitted to future generations. The golden rule in this sequential game should be to bequeath to future generations a state of atmospheric GHG concentrations and projected emissions from existing physical capital that enables them to avoid 'dangerous interference with the climate system' without having to make a disproportionate sacrifice.

NOTES

1. Nicholas Georgescu-Roegen launched the theme of 'degrowth' in the early 1970s (see his seminal book *The Entropy Law and the Economic Process*, 1971). This subject has enjoyed a revival since the beginning of the twenty-first century with the publication of several works, including notably that of Serge Latouche (2006), *Le Pari de la décroissance*.
2. See the works of Amartya Sen for a powerful reflection on the invalidity of the sequence 'utility–preferences of the individual–exercise of choice–market values' and his reverse interpretation, working up from market values to utility (see, for example, Sen, 1999).
3. The theory of this curve was that in its initial stages, the process of growth exacerbated environmental destruction, but that the same process of growth subsequently reduced its environmental impact. Whence the idea of accelerating growth to attain this 'best of all worlds' as soon as possible! This theory has never been confirmed empirically, except by an *ad hoc* selection of the environmental parameters considered, and even less so for the problem of climate change than for local environmental problems, such as water management or urban air pollution (see Cole et al., 1997, and Stern, 2004).
4. In particular, this article, 'Is there a need for a new environmental ethics?', published in the Proceedings of the XV World Congress of Philosophy, included Sylvan's 'last man' argument: if the last man on Earth, who knows he is about to die, sets about destroying all other life on Earth, is there anything ethically *wrong* with that, seeing as how it will never be of any utility to other humans? If our moral intuition leads us to answer 'yes', then it must be because we recognize intrinsic value in other living beings, even those that do not possess 'conscience' or sensibility (Sylvan, 1973).
5. Naess coined the expression 'deep ecology', intended to mark a break with superficial, management-oriented ecology, which remained anthropocentric (Naess, 1973).
6. Callicott extended the work of Leopold on 'land ethics' by seeking wider philosophical foundations. In the mid-1980s, however, Callicott came to the conclusion that the idea of intrinsic value could not exist objectively (Callicott, 1979).
7. This section is based on Godard (2006).
8. The basic principle of Leopold's 'land ethic' is that 'a thing is right when it tends to preserve the integrity, stability and beauty of the biotic community. It is wrong when it tends otherwise' (Leopold, 1949 [1989], pp. 224–5).
9. I owe this expression to the biologist Francisco Varela (1979).
10. This is why one can represent the organization of an autonomous system as a network of components and processes that are all looped together in such a way as to ensure permanence over time, despite fluctuations in the environment and the renewal of its physical components.
11. This idea of hierarchy in sources of meaning can be understood from the relation > signifying '. . . proceeds from . . .'. If that which constitutes Y as an entity proceeds from X, then we say that X is hierarchically superior to Y: $X > Y$.
12. This concept was introduced by the cognitive science theorist Douglas Hofstadter (1979).
13. For neoclassical theory, a so-called 'technological' externality designates an interaction between the utility or objective functions of two or more agents that does not take the form of a voluntary transaction in the market and does not, therefore, benefit from the regulation of behaviour provided by the price system. The existence of such

externalities introduces a gap between the costs actually incurred by the agents producing the externality and the total costs resulting from their activity, part of which are passed on to other agents or to the collectivity as a whole. Neoclassical theory sees in this the source of an economic inefficiency to be corrected. Note that the 'external' nature of these effects does not refer to an action to the natural environment, but simply to the off-market and economically unregulated character of the relationship between agents.

14. Whence Leopold's phrase, quoted as an epigraph to this chapter: 'Thinking like a mountain' (Leopold, 1949 [1989]).
15. This idea of sustainable development was popularized in 1987 by the report of the World Commission on the Environment and Development (WCED), called the 'Brundtland Report' after its chairwoman. The commission was convened by the United Nations, and its report contains the now-famous phrase: 'Sustainable development is development that meets the needs of the present without compromising the ability of future generations to meet their own needs' (WCED, 1987). In particular, this phrase has directly inspired French legislation.
16. Pearce's model has provoked fierce criticism, pointing out the contradiction between his premises (the standard form of the external damage curve) and the apocalyptic outcome or, on a more technical level, the unsuitability of a static frame of analysis for tackling a dynamic problem. As a consequence, models developed since then have most often adopted the formalism of the dynamic theory of optimal control. In fact, however, these criticisms are unconvincing on the most important points: (a) whatever the form of the external damage curve, it can only come into being, by construction, when the threshold of assimilation capacity has been exceeded, meaning that the optimum of neoclassic internalization necessarily entails a certain level of net destruction of this capacity, once the rate of natural restoration has been deducted; (b) the treatment of the dynamics of choices as a succession of decision problems defined statically echoes the sequential approach recommended nowadays for dealing with the dynamics of uncertainty; and (c) this type of modelling has great heuristic relevance for explaining the way in which public decision makers approach problems of decision. On the sequential approach, see Hourcade (1997).
17. In the context of flow pollutant models, see, for example, Leandri (2009).
18. A discount rate that many economists considered to be excessively low: 1.4 per cent per year, determined essentially on the hypothesis of long-term, per capita economic growth (1.3 per cent).
19. Nordhaus used a rate of 5.5 per cent (see Nordhaus, 2007).
20. In a growing economy, it is not enough for technical progress to improve the environmental performance per unit of product; the total environmental impact of human activity must be reduced from one period to the next to keep in line with the falling carrying capacity of the environment.
21. In fact, this revision is already well under way, particularly in economic works on sustainable development, which build on the concept of natural capital, but it is not always recognized as such. Whence the feeling of divergence between the content of some of the particular models proposed and the theoretical foundation on which they are supposed to be built.
22. Not all externalities are of the sort considered in this text. Ordinary noise pollution, for example, affects human welfare but does not alter basic environmental functions.
23. The geophysicist David Archer concluded from his modelling of atmosphere–ocean–land biosphere interactions that it would take about 400,000 years to return to the pre-industrial atmospheric GHG concentration of 280 ppm. His calculations show that a significant proportion (25 per cent) of the CO_2 emitted by humanity must be considered to remain in the atmosphere practically eternally. Whence an estimation of the average life of atmospheric CO_2 of 35,000 years, or to put it another way: 'an average life of 300 years for 75% of emissions and almost-eternity for the other 25%' (Archer, 2005).

REFERENCES

Afeissa, H.S. (ed.) (2007), *Éthique de l'environnement – Nature, valeur, respect*, Paris: Vrin (Coll. Textes clés).

Archer, D. (2005), 'Fate of fossil fuel CO_2 in geologic time', *Journal of Geophysical Research*, **110**, C09S05; doi: 10.1029/2004JC002625.

Arrow, K. and A.C. Fisher (1974), 'Environmental preservation, uncertainty and irreversibility', *Quarterly Journal of Economics*, **88**, pp. 312–19.

Auerswald, P., L.M. Branscomb, T.M. La Porte and E.O. Michel-Kerjan (eds) (2006), *Seeds of Disaster, Roots of Response*, Cambridge: Cambridge University Press.

Ayres, R.U. (1994), 'Industrial metabolism: theory and policy', in R.U. Ayres and U.K. Simonis (eds), *Industrial Metabolism: Restructuring for Sustainable Development*, Tokyo: United Nations University Press, pp. 3–20.

Ayres, R.U. and A.V. Kneese (1969), 'Production, consumption and externalities', *American Economic Review*, **59**, pp. 282–97.

Baumol, W.J. and W.E. Oates (1975), *The Theory of Environmental Policy: Externalities, Public Outlays and the Quality of Life*, Englewood Cliffs, NJ: Prentice-Hall.

Beck, U. (1992), *Risk Society: Towards a New Modernity*, London: Sage.

Beckerman, W. (1974), *In Defence of Economic Growth*, London: Jonathan Cape.

Boltanski, L. and L. Thévenot (1991), *De la justification. Les économies de la grandeur*, Paris: Gallimard (Coll. NRF – Les Essais). English edition: Boltanski, L. and L. Thévenot (2006), *On Justification – Economies of Worth*, Princeton, NJ: Princeton University Press.

Boulding, K.E. (1966), 'The economics of the coming spaceship Earth', in H. Jarrett (ed.), *Environmental Quality in a Growing Economy*, Baltimore, MD: Johns Hopkins University Press, pp. 3–14.

Bromley, D.W. (ed.) (1995), *The Handbook of Environmental Economics*, Oxford: Blackwell.

Callicott, J.B. (1979), 'Elements of an environmental ethic: moral considerability and the biotic community', *Environmental Ethics*, **1**, pp. 71–81.

Castoriadis, C. (1975), *L'institution imaginaire de la société*, Paris: Seuil. English edition: Castoriadis, C. (1998), *The Imaginary Institution of Society*, Cambridge, MA: MIT Press.

Coase, R. (1960), 'The problem of social cost', *Journal of Law and Economics*, **3**, pp. 1–44.

Cole, M.A., A.J. Rayner and J.M. Bates (1997), 'The environmental Kuznets curve: an empirical analysis', *Environment and Development Economics*, **2** (4), pp. 401–16.

Costanza, R. (ed.) (1991), *Ecological Economics: The Science and Management of Sustainability*, New York: Columbia University Press.

Cropper, M.L. and W.E. Oates (1992), 'Environmental economics: a survey', *Journal of Economic Literature*, **30** (2), pp. 675–740.

Daly, H.E. (1968), 'On economics as a life science', *Journal of Political Economy*, **76**, pp. 392–406.

De Perthuis, C. (2003), *La Génération future a-t-elle un avenir? Développement durable et mondialisation*, Paris: Belin.

Elliott, D. and G. Yarrow (1977), 'Cost–benefit analysis and environmental policy: a comment', *Kyklos*, **30** (2), pp. 300–309.

Faber, M., H. Niemes and G. Stephan (1987), *Entropy, Environment and Resources – An Essay in Physico-Economics*, Berlin: Springer Verlag.

Farvar, T. and J.P. Milton (eds) (1972), *The Careless Technology. Ecology and International Development*, New York: Natural History Press.

Forrester, J.W. (1971), *World Dynamics*, Cambridge, MA: Wright-Allen Press.

Foster, J. (ed.) (1997), *Valuing Nature? Economics, Ethics and Environment*, London: Routledge.

Georgescu-Roegen, N. (1971), *The Entropy Law and the Economic Process*, Cambridge, MA: Harvard University Press.

Godard, O. (1990), 'Environnement, modes de coordination et systèmes de légitimité : analyse de la catégorie de patrimoine naturel', *Revue économique*, **41** (2), pp. 215–41.

Godard, O. (1997a), 'Le concept d'environnement, une hiérarchie enchevêtrée', in C. Larrère and R. Larrère (eds), *La crise environnementale*, Paris: INRA-Éditions (Coll. Les colloques No. 80), pp. 97–112.

Godard, O. (1997b), 'Social decision-making under scientific controversy, expertise and the precautionary principle', in C. Joerges, K.H. Ladeur and E. Vos (eds), *Integrating Scientific Expertise into Regulatory Decision-making: National Experiences and European Innovations*, Baden-Baden: Nomos Verlagsgesellschaft, pp. 39–73.

Godard, O. (1998), 'L'écodéveloppement revisité', *Économies et sociétés*, 'Développement, croissance et progrès', série F. **36** (1), pp. 213–29.

Godard, O. (2003), 'Y a-t-il des limites à l'utilisation de la nature?', in C. Lévêque and S. van der Leeuw (eds), *Quelles natures voulons-nous? Pour une approche socio-écologique du champ de l'environnement*, Paris: Elsevier, (Coll. Environnement), pp. 197–207.

Godard, O. (2004), 'De la pluralité des ordres – Les problèmes d'environnement et de développement durable à la lumière de la théorie de la justification', *Géographie, Économie, Société*, **6** (3), pp. 303–30.

Godard, O. (2006), 'La pensée économique face à l'environnement', in A. Leroux and P. Livet (eds), *Leçons de Philosophie économique – Tome II : Économie normative et philosophie morale*, Paris: Economica, pp. 241–77.

Godard, O. (2007a), 'Le Rapport Stern sur l'économie du changement climatique était-il une manipulation grossière de la méthodologie économique?', *Revue d'économie politique*, **117** (4), pp. 479–511.

Godard, O. (2007b), 'Peut-on séparer de façon précoce le bon grain de l'ivraie?', in C. Kermisch and G. Hottois (eds), *Techniques et philosophie des risques*, Paris: Vrin (Coll. Pour demain), pp. 139–57.

Godard, O. (2008), 'The Stern Review on the Economics of Climate Change: contents, insights and assessment of the critical debate', *Surveys and Perspectives Integrating Environment and Society* (S.A.P.I.E.N.S), **1** (1), pp. 23–41, available at: http://sapiens.revues.org/index240.html.

Godard, O., C. Henry, P. Lagadec and E. Michel-Kerjan (2002), *Traité des nouveaux risques. Précaution, crise et assurance*, Paris: Gallimard (Coll. Folio-Actuel 100).

Gordon, H.S. (1954), 'The economic theory of a common property resource: the fishery', *Journal of Political Economy*, **62** (2), pp. 124–42.

Hansen, J., M. Sato, P. Kharecha, D. Beerling, R. Berner, V. Masson-Delmotte, M.

Pagani, M. Raymo, D.L. Royer and J.C. Zachos (2008), 'Target atmospheric CO_2: where should humanity aim?', *Open Atmospheric Science Journal*, **2**, pp. 217–31.

Henry, C. (1974), 'Investment decisions under uncertainty: the irreversibility effect', *American Economic Review*, **64**, pp. 1006–12.

Hofstadter, D. (1979), *Gödel, Escher, Bach: An Eternal Golden Braid*, New York: Basic Books.

Hotelling, H. (1931), 'The economics of exhaustible resources', *Journal of Political Economy*, **39** (2), pp. 137–75.

Hourcade, J.C. (1997), 'Précaution et approche séquentielle de la décision face aux risques climatiques de l'effet de serre', in O. Godard (ed.), *Le principe de précaution dans la conduite des affaires humaines*, Paris: Éd. de la MSH et INRA-Éditions, pp. 259–94.

Jonas, H. (1984), *The Imperative of Responsibility: In Search of an Ethics for the Technological Age*, Chicago, IL: University of Chicago Press. French edition: Jonas, H. (1990), *Le principe responsabilité. Une éthique pour la civilisation technologique*, Paris: Cerf.

Kapp, K.W. (1950), *The Social Costs of Private Enterprise*, Cambridge, MA: Harvard University Press.

Kneese, A. (1964), *The Economics of Regional Water Quality Management*, Baltimore, MD: Johns Hopkins University Press.

Larrère, C. (1997), *Les Philosophies de l'environnement*, Paris: PUF (Coll. Les philosophes).

Latouche, S. (2006), *Le Pari de la décroissance*, Paris: Fayard.

Le Treut, H. (2004), 'La base scientifique', in H. Le Treut, J.-P. van Ypersele, S. Hallegatte and J.-C. Hourcade (eds), *Science du changement climatique – Acquis et controverses*, Paris: IDDRI, pp. 11–29.

Leandri, M. (2009), 'The shadow price of assimilative capacity in optimal flow pollution control', *Ecological Economics*, **68** (4), pp. 1020–31.

Leopold, A. (1949 [1989]), *A Sand County Almanac and Sketches Here and There*, Oxford: Oxford University Press.

Majone, G. (1989), *Evidence, Argument and Persuasion in the Policy Process*, New Haven, CT and London: Yale University Press.

Malinvaud, E. (1975), *Leçons de théorie microéconomique*, 3rd edn, Paris: Dunod.

Malthus, T.R. (1803), *An Essay on the Principle of Population*, 2nd edn, London: Johnson.

Meadows, D.H. and D.L. Meadows, J. Randers and W.W. Behrens III (eds) (1972), *The Limits to Growth. A Report to the Club of Rome*, New York: Universe Books.

Mishan, E. (1967), *The Costs of Economic Growth*, London: Staples Press.

Naess, A. (1973), 'The shallow and the deep, long-range ecology movements', *Inquiry*, **16** (1), pp. 95–100.

Nordhaus, W.D. (2007), 'A review of *The Stern Review on the Economics of Climate Change*', *Journal of Economic Literature*, **45** (3), pp. 686–702.

Oates, W.E. (ed.) (1994), *The Economics of the Environment*, Aldershot, UK and Brookfield, VT, USA: Edward Elgar.

Passet, R. (1979), *L'économique et le vivant*, Paris: Payot (2nd edition 1996, Economica).

Pearce, D.W. (1976), 'The limits of cost–benefit analysis as a guide to environmental policy', *Kyklos*, **29** (1), pp. 97–112.

Pigou, A.C. (1920), *The Economics of Welfare*, London: Macmillan.

Ricardo, D. (1817), *On the Principles of Political Economy and Taxation*, London: John Murray (3rd edn, 1821).

Rostow, W. (1960), *The Stages of Economic Growth: A Non-Communist Manifesto*, Cambridge: Cambridge University Press.

Sachs, I. (1972), 'Environnement et projet de civilisation', *Les Temps modernes*, **28** (316), pp. 1–14.

Sachs, I. (1974), 'Environnement et styles de développement', *Les Annales. Économies, Sociétés Civilisations*, **3**, pp. 553–70.

Sachs, I. (1980), *Stratégies de l'écodéveloppement*, Paris: Éditions Ouvrières and Économie et Humanisme.

Sachs, I. (1998), *L'écodéveloppement. Stratégies pour le XXIe siècle*, Paris: Syros-La Découverte.

Sagoff, M. (1988), *The Economy of the Earth: Philosophy, Law, and the Environment*, Cambridge and New York: Cambridge University Press.

Samuelson, P. (1954), 'The pure theory of public expenditures', *Review of Economics and Statistics*, **36** (4), pp. 387–9.

Scitovsky, T. (1955), 'Two concepts of external economies', *Journal of Political Economy*, **62** (2), pp. 143–51.

Sen, A. (1999), *Development as Freedom*, New York: Alfred Knopf.

Shaefer, M.B. (1955), 'Some considerations of population dynamics and economics in relation to the management of the commercial marine fisheries', *Journal of the Fisheries Resources Board of Canada*, **14** (5), pp. 669–81.

Shrader-Frechette, K.S. (1981), *Environmental Ethics*, Pacific Grove, CA: Boxwood Press.

Solow, R.M. (1974), 'The economics of resources or the resources of economics – Richard T. Ely Lecture', *American Economic Review*, **64** (2), pp. 1–14.

Solow, R.M. (1993), 'An almost practical step toward sustainability', *Resources Policy*, **19** (3), pp. 162–72.

Stern, D.I. (2004), 'The rise and fall of the environmental Kuznets curve', *World Development*, **32** (8), pp. 1419–39.

Stern, N. (ed.) (2006), *The Stern Review on the Economics of Climate Change*, London: HM Treasury, 30 October.

Sterner, T. and J. van den Bergh (eds) (1998), 'Special issue: frontiers of environmental and resource economics: testing the theories', *Environmental and Resource Economics*, **11** (3–4), pp. 243–60.

Sylvan, R. (1973), 'Is there a need for a new environmental ethics?', in *Philosophy and Science – Morality and Culture – Technology and Man –Proceedings of the XV World Congress of Philosophy*, Vol. 1. Varna (Bulgaria), pp. 205–10.

Varela, F. (1979), *Principles of Biological Autonomy*, New York: Elsevier/North-Holland. French edition: Varela, F. (1989), *Autonomie et connaissance*, Paris: Seuil (Coll. La Couleur des idées).

WCED (1987), *Our Common Future*, Oxford: Oxford University Press.

Weisbrod, B. (1964), 'Collective-consumption services of individual-consumption goods', *Quarterly Journal of Economics*, **78** (3), pp. 471–7.

Weitzman, M.L. (2007), 'A review of *The Stern Review on the Economics of Climate Change*', *Journal of Economic Literature*, **45** (3), pp. 703–24.

Weitzman, M.L. (2009), 'On modeling and interpreting the economics of catastrophic climate change', *Review of Economics and Statistics*, February, **91** (1), pp. 1–19.

3. Building scenarios: how climate change became an economic question

Michel Armatte

INTRODUCTION

My position in the debate on climate change is not principally that of an economist involved in model-based research on the human forcing of climate change, the scale of its impact on societies or the optimization of policies of mitigation, adaptation and innovation. I am interested in such research from the reflexive viewpoint of the history and sociology of sciences, or of science studies in general knowledge. Such a position has the disadvantage of not making direct statements about climate change, but has the advantage of not being restricted to any particular one of the scientific communities acting in this domain.

My reflection on these subjects started with questions about modelling and its origins in the different traditions of meteorologists and economists (Armatte and Dahan Dalmedico, 2004). Questions arose about limited growth in the 1970s,[1] challenging models of the climate system (in the tradition of models of atmospheric circulation and its coupling with oceans and vegetation) and pathways of greenhouse gas (GHG) emissions and their impact in integrated assessment models (IAMs), which help to refine cost–benefit calculations for policies of mitigation.

I then became more interested in the complementary approach of scenarios, the roots of which can be traced back to the 'prospective method' of Cold War think-tanks. For France, the approach began with the pioneering work of Gaston Berger and Bertrand de Jouvenel, moving on to the indicative planning of the *Commissariat au Plan*, the Délégation à l'aménagement du territoire et à l'action régionale (DATAR), and the large state-owned companies. Considering a range of scenarios appeared to be the only way to take into account the variety of possible futures, ranging from the conservative – such as the 'business-as-usual' scenario – to those resulting from pro-active policies of reduction, which may be probable, improbable or even extreme. Discussing the forms that these scenarios could take, their mode of construction and their role in coherent descriptions of the future, either in exploratory form – moving forward from the present – or in the form of more directly forecast futures, I followed the vagaries of their implementation in projects such as the 'Scenario of the Unacceptable' (DATAR, 1971), 'Interfuture' (OECD, 1979) or 'Scanning the Future' (CPB, 1992) (see Armatte 2007, 2008). A scenario can therefore be defined, as the authors of the *Special Report on Emissions Scenarios* (SRES) state, as 'A plausible description of how the future may develop, based on a coherent and internally consistent set of assumptions' (Nakicenovic et al., 2000). It can take one of two very complementary forms: that of a storyline describing the future world and the evolutions that lead to it, or that of an IAM, quantifying the hypotheses of the scenario (in terms of population, growth, energy contents, trade and so on) and formalizing their synchronic links and dynamics.

The Intergovernmental Panel on Climate Change (IPCC) decided to use this method as early as 1992, but during the present decade, the SRES has defined a simplified framework, validated by the Working Group III, of about 40 scenarios, divided into four families. These four families are defined by crossing two sets of divergent tendencies, one set contrasting unsustainable growth with sustainable growth, and the other contrasting increasing globalization with increasing regionalization. Each of these four basic scenarios corresponds to a set of contrasting hypotheses about the main factors of climate forcing.

How are these scenarios, and indeed the scenario method itself, appreciated today, in the present climate context? To answer this question, I have plunged into the Fourth Assessment Report of the IPCC (2007a, 2007b).

THE SPECIAL REPORT ON EMISSIONS SCENARIOS IN THE LIGHT OF THE IPCC FOURTH ASSESSMENT REPORT

This fourth report confirms the role of the SRES as a basis for emissions simulations. The definitions from the Third Assessment Report are repeated word for word and enshrined in the form of explicit inserts.[2] The 'Summary for Policymakers' drafted by Working Group I draws on the main families of scenarios to establish the 'best estimates' of average temperature increases ranging from 0.6°C to 4°C. The figures for each scenario have since been revised for the new simulations required for the Fourth Assessment Report. The climatologists of Working Group I carried out the simulations requested by the IPCC in the context of the scenarios proposed. It appears that the exercise caused them some problems[3] however, for the IPCC's framing of their research has both advantages and drawbacks. On the positive side, these scenarios enable them to compare different climate simulations for the same forcing conditions. The downside is that they are required to calibrate governance modelling based on soft objects such as the scenarios, which are, after all, no more than stories of possibilities. The climatologists, however, would rather concentrate their efforts on calibrating the scientific research of unsolved physical problems. Although the climatologists acknowledge the advantages of being able to compare the results of several dozen atmosphere-ocean general circulation-type climate models for a same kind of scenario, breaking up the high uncertainty attached to predictions, they find that the drawbacks still outweigh the benefits. They have also expressed feelings of being exploited by this vast benchmarking exercise, as they are no longer able to choose their own working hypotheses and priorities, which are now often regulated and determined by the needs of the IPCC reports.

The 2007 report by Working Group II on climate change impacts, adaptation and vulnerability (IPCC 2007a) provides a more complete assessment of the approach to impacts conditioned by the scenarios and by integrated assessment. The reviews of the Millennium Ecosystem Assessment (Alcamo et al., 2005) and the Global Scenarios Group (Raskin, 2005) bear witness to the uses of scenarios. The report stresses two recent trends in the modelling of impacts. The first is that of high resolution (50 km), which, compared to the first-generation general circulation models, enables impact scenarios to be drawn up on a local scale, which is more efficient than a simple local variation of the main socio-economic aggregates (population, income) or the 'statistical downscaling' sometimes practised. The second trend involves treating the question of impacts and responses more as a problem of management and decision making than as a problem of knowledge. The analysis of risks is then no more than one moment in a global risk management approach combining consideration of the costs' possible impacts with the probability of their occurrence. Nevertheless, this irruption of probability has also raised serious problems (see below).

Working Group III was given the task of evaluating policies (mitigation, adaptation and innovation). The report – written by Nebojsa Nakicenovic et al., the authors responsible for the SRES – presents a view that differs little from the initial concept of scenario. For these authors, the idea of scenario, in its very nature as a tool for exploring alternative futures, remains totally antinomic to the idea of deterministic or probabilistic projection. It excludes, however, both 'catastrophist' scenarios (such as the Pentagon's story of the inversion of the Gulf Stream) and overoptimistic scenarios of a future solution in the form of technological innovation. Policies were based on two sorts of scenarios: ones that define a target (for example, a level of CO_2 concentration for 2100), and ones that define policy measures for reducing emissions, including the *laissez-faire* approach: 'baseline' or 'business as usual' scenarios. The report lists no less than 800 emission scenarios used in the different studies, of which 400 are of the baseline type. These baseline scenarios are usually founded on the Kaya identity, which breaks emissions down into four factors: population, per-capita production, energy intensity of production, and carbon intensity of energy consumed. The level of activity – in other words growth – remains the main factor of GHG emissions, and modelling it at the global scale constitutes, together with population projections, the basis of an emissions scenario.[4] In this respect, Working Group III's contribution to the fourth report appears to be a simple readjustment of the SRES in the light of the studies of the six preceding years, with the correction of demographic and energy trends, a better assessment of the role of land cultivation and the taking into account of other GHGs and pollutants. For the reporters, the wide diversity of scenarios exploring the feasibility and cost of mitigation policies is mainly due to imprecision about the goal to be achieved. The United Nations Framework Convention on Climate Change (UNFCCC) defined it as 'stabilising the atmospheric concentrations of greenhouse gases', but without specifying at what level, and the studies are spread over more or less drastic objectives. These objectives may concern CO_2 alone, or all greenhouse gases, and this entails choices of substitution. They may concern GHG concentrations or radiative forcing – two different indicators that are, fortunately, relatively well correlated – and the goals may have fixed or moving horizons.

Abandoning an a priori classification of scenarios according to their socio-demographic and economic characteristics, the authors of the fourth report proposed an a posteriori classification of the 150-odd scenarios of mitigation drawn from the literature into six categories depending on the targets expressed in terms of concentration, emissions and temperature increase. Moss et al. (2008, p. 103) organize their data according to effects, and no longer according to their causes (or 'driving forces').

Ten years of use in the study of climate change has therefore not exhausted the concept of scenario. A reading of the 2007 reports of the three IPCC working groups, however, brings out numerous dissatisfactions with the precise role attributed to them during this decade.

● The SRES scenarios were not drawn up with either the necessary transparency or the support of the whole scientific community.

- The huge number of variations on the scenarios has become counterproductive: it makes the simulations incommensurable and incomprehensible for policy makers.
- The simulation of business-as-usual-type scenarios appears to dominate and to distort the idea of the multiplicity of possible futures to be envisaged, including the most extreme cases.
- The fact that other scenarios adopt, on the contrary, prospects of highly oriented political intervention results in a subordination of scientific research to politics. This has frustrated some physicists and biologists and runs counter to the usual linear process of expertise.
- Climatologists have suffered from being too dependent on the IAM community for the forcing scenarios that they have to illustrate in their simulation, including in terms of timetables. Moreover, the imperatives of writing the reports made it impossible to prolong or multiply these simulations.
- Globalization and sustainable development, on which the four families A1, A2, B1 and B2 are based, no longer have the same topicality and content as they had in the 1980s. As possible strategies, globalization and sustainable development are no longer on the same level of likelihood as their opposites: regionalization, protectionism and strong growth based on fossil fuels. They have become shared and almost inevitable options. Doubtless, the departure of the Bush administration and the context of the economic crisis will once again change not only the relative levels of globalization and sustainability, but also their content.
- The balance between development and protection of the environment, which is not arbitrated in the same way in the North as in the South, was not taken into account in 2000. It assumes greater importance today. The role of developing countries in drawing up the SRES scenarios has been particularly insufficient, both as economic actors and as authors of their own strategic scenarios. The choice between short and long term, which cannot be reduced to a simple discount factor, and the question of inequality between regions in their exposure to climate risk was not given the attention it deserved in the creation of the scenarios of 2000.
- Lastly, the demand for forecasts on a local scale is starting to eclipse the almost exclusive interest in global climate change and its expression in one sole quantity: global average surface temperature. The differing vulnerability to climate change of different regions and micro-regions may well prove to be more important than the average temperature increase of the whole planet.

TOWARDS NEW SCENARIOS: THE REFORM

Following several meetings on scenarios in July 2005 at the International Institute for Applied Systems Analysis (IIASA) of Vienna[5] and in March 2006 in Seville, the decision to revise the scenarios adopted by the SRES was made during the 25th IPCC session on Mauritius in April 2006. It was decided that the revision would be carried out by a committee of experts representing the research community and given the mission of identifying a set of benchmark concentration scenarios 'compatible with the full range of emission stabilization and baseline scenarios currently available in the scientific literature'. These new scenarios were drawn up during meetings of the different modelling organizations[6] and a preparatory document, the Background Paper, was produced, written under the initiative of Richard Moss et al.[7] This document brought together the material submitted to a group of 130 experts, who met on 19, 20 and 21 September 2007 in Noordwijkerhout, Netherlands, to establish the new scenarios. These experts included modellers from the following three communities: integrated assessment modelling (IAM), impacts, adaptation and vulnerability (IAV) and climate modelling (CM), considered as 'intermediate users', together with UNFCCC negotiators and non-governmental organizations (NGOs), considered as the final users. Contrary to the drafting of the SRES, which was coordinated and voted by the IPCC, and which thus imposed a set of highly structured scenarios from outside the scientific and political communities, this group of experts was intended to represent a direct emanation of the users, with the role of the IPCC reduced to that of catalyst.

The reworking of scenarios carried out by this group, now the subject of a special report – *Towards New Scenarios* (TNS: Moss et al., 2008) – was based more on the function of scenarios in the process of evaluation and less on their content, in the sense of their constituent hypotheses. The elements shaping this reconstruction went beyond simple cognitive rationality: they had become strategic. Two main series of imperatives had to be satisfied.

First, were imperatives resulting from a social demand that was much more pressing than it had been 10 years earlier:

- distributing time scales – associating long-term targets with pathways enabling medium-term policies to be implemented as from now;
- distributing geographical scales – knowing how to present the impact of emissions at various regional and local scales;
- taking all factors into account (the different GHGs, the carbon cycle, plant cover and so on) and the feedbacks between these factors;
- expressing the majority of uncertainties in terms of risks.

Second, organizational imperatives resulting from needs expressed by the scientific communities for better interaction between their research:

- finding a minimal shared base of scenarios accepted by everyone;
- giving more autonomy and time to climatologists in relation to the economists producing emissions scenarios.

The solution proposed by the group of experts to these two sets of constraints on new scenarios, developed during the meetings in Vienna, Seville, Mauritius and so on, consisted of three main moves.

The first move was to define two different horizons for the new scenarios.

- The near term, focusing on the date 2035, is intended to enable more precise identification of the risks (costs and probabilities) of isolated or repeated climate events, better assessment of the impacts and the vulnerability of populations and economic and biological systems, and clearer knowledge of the possibilities of adaptation, with a view to preparing responses in terms of investment and regulation. At this horizon, it is possible to develop high-resolution modelling (in terms of both time and space) that greatly reduces uncertainties.
- The long term remains just as necessary for defining major and rare risks, identifying thresholds, and choosing strategies of mitigation, adaptation and development that are robust with regard to the climatic and social uncertainties that affect the near term. The long term can also take into account the huge inertia of certain phenomena such as the response of deep oceans and ice caps. Two horizons were chosen for the long-term studies: 2100 and 2300.

The second move was to replace the sequential approach that governed the previous two reports (AR3 and AR4) by a parallel approach.

The sequential approach, summarized in Box 3.1, took a SRES-type emissions scenario as the point of departure of the assessment process. That scenario served as the basis for establishing the radiative forcing data for a general circulation model producing climate projections for 2050 or 2100. Researchers in Group III then used these projections to draw up impact profiles and assess the vulnerability of subsystems to those impacts, along with their capacity for adaptation or defence. This 'driving forces-pressures-states-impacts-responses' type of analytic framework was visible in the very organization of the IPCC into three groups, and even more so in the structure of most of the integrated models, such as the IMAGE 2.2, which presented the complete range of projection indicators according to this framework.

BOX 3.1   APPROACHES TO THE DEVELOPMENT OF GLOBAL SCENARIOS

The *sequential approach* is linear, hierarchical and unilateral. It begins with emissions and socio-economic scenarios (IAMs), continues with radiative forcing, followed by climate projections, ending with impacts, adaptation and vulnerability (IAV).

The proposed *parallel approach* breaks down the linear, hierarchical sequential approach into a circular one. The process now begins with the representative concentration pathways (RCPs) and levels of radiative forcing. The second step is in two, separate parts: (i) climate, atmospheric and C-cycle projects (CMs) and (ii) IAMs. These steps not only occur concomitantly, but also share and integrate information between themselves. This is followed by IAV and mitigation analysis, which integrates information and sends feedbacks to (i) and (ii). This communicative process can alter the choice of RCPs, leading back to the first step.

These approaches are elaborated in Figure 1 of Moss et al. (2008, p. iv).

The TNS experts have envisaged a process in which the two CM and IAM communities work in parallel, one on atmosphere-ocean general circulation models and earth system models, and the other on projections of the socio-economic activities producing GHG emissions scenarios. The two communities now use a small number of representative concentration pathways (RCPs) – originally called benchmark concentration pathways – which can be interpreted as traces in the atmosphere of the scenarios of climate forcing. That means that the climatologists no longer have to wait for the definition or re-definition of emissions scenarios by economists before running or re-running simulations of atmosphere and ocean circulation. They thus no longer need the detailed socio-economic characteristics that condition emissions. They can now base their calculations directly on the level of climate forcing given by the chosen RCP, and because of this, they are able to achieve higher definition (mesh sizes of 0.5°C for the near term and 2°C for the long term) and greater realism, taking into account new factors such as cloud cover, the salinity of the oceans, the variability of plant cover, carbon biochemistry and cycles. In this way, their general circulation-type models are developing towards large earth system models that integrate these different elements of knowledge better than the techniques of coupling and parameterization could.[8] Likewise, economists of the IAM community who have provided the bases for these RCPs in the past will now be able to explore a large number of possible scenarios without troubling their neighbours, as long as they can attribute a result in terms of RCP to each of them.

This is because each of the RCPs adopted is *representative* of at least one and usually several emissions scenarios that have already been studied and that produce the same radiative forcing of the climate system. The experts wanted RCP to be 'compatible with the full range of emission stabilization and baseline scenarios currently available in the scientific literature'. Starting with the six classes of scenarios found in the Fourth Assessment Report of the IPCC (see above), the experts at the New Scenario Meeting agreed on the figure of four RCPs. Two extreme pathways, referred to as the low and the high RCPs, represent the first and last deciles of the 300 scenarios studied in the post-SRES literature. Two others represent intermediate scenarios. The low pathway is the only one of the 'overshoot' and 'peak-and-decline' form, which is to say that it reaches a peak of $3W/m^2$ in about 2040–50 and then falls back towards $2.6 \ W/m^2$. The experts were very interested in such a virtuous profile. The high pathway, roughly corresponding to a carbon-intensive SRES A2 scenario (but revised), grows steadily up to $8.5 \ W/m^2$ in 2100 and then continues to increase after that date. Intermediate profiles were useful for understanding the non-linearities caused by relations of interdependence and feedback between demographic, economic, energy and climatic factors. The choice, however, of one sole intermediate RCP would have turned it into a main reference, or even a norm. Two intermediate RCPs were therefore chosen with values of $4.5 \ W/m^2$ and $6 \ W/m^2$ (Moss et al., 2008, p. 34).

The third move was to choose good 'representatives' for these four RCPs, in other words to choose the team and model that would embody each of them. Twenty IAMs from seven teams were identified as candidates. The Steering Committee of the group of experts chose four representatives: MESSAGE for RCP8.5, AIM for RCP6, MiniCAM for RCP4.5 and IMAGE for RCP3 (Moss et al., 2008, p. 38).

For each of the RCP candidates, it was possible to associate a profile of radiative forcing (falling between 3 and $8.5 \ W/m^2$) and a profile of CO_2 emissions produced by the industrial and energy sectors (ranging from about 0 to 100 gigatons in the year 2100), from among the profiles presented in Figure 5 of the report, on which the four models chosen as representatives can be situated. The choice of IMAGE 2.6 rather than the more moderate IMAGE 2.9 as representative of the RCP3 sparked an interesting controversy about whether or not it was necessary to adapt to recent proposals by the G8 to cut emissions by 50 per cent by 2050, or by the European Union to adopt the goal of a 2°C temperature increase (that is a GHG concentration of about 400 parts per million (ppm) and a forcing of 2W/m²). For some authors, not including the most drastic scenarios, pathways and policies of mitigation considered in the negotiations and scientific literature was in itself 'policy prescriptive' rather than 'policy relevant'.

THE SIGNIFICANCE OF THE REFORM

The Scenario as a Tool for Organizing Work between Communities

The reworking of the concept of scenario is anchored in a resolutely socio-cognitive view of climate sciences, in the sense that it puts into practice a view, or even a paradigm, that never dissociates the scientific aspects from the socio-political ones. According to this view, it must be specified how to grasp, describe ever more precisely and explore dynamically all the possible futures of the Earth's system – a more and more complex system in the sense that it incorporates subsystems that are very different in essence (atmosphere, oceans, forests, biotopes and so on), indistinctly natural and anthropogenic, and with very different temporal and spatial scales. It must also be specified, however, how to organize the action, in other words how to get the scientists, experts and policy makers to work together.

The reform of the scenarios thus seeks to answer the needs of communities in terms of either the production of assessments or the negotiation of policies of mitigation and adaptation. It lies within the scope of the division of labour between these communities, mainly in response to a demand for policy assessment, which prevails over purely epistemological rationales and over the organization of the supply of scenarios, models and results to meet that demand. 'The parallel process described in this document provides a strategy for explicit engagement between the communities', the report states (Moss et al., 2008, p. xxi). It is a new contract.

The contractual commitment applies, first, between the scientists and the policy makers interested in the results of the studies. In addition to the climate negotiators, the IPCC projections are of great interest to a wide range of NGOs and UN organizations working in the domains of development or the environment, such as the Food and Agriculture Organization (FAO) of the United Nations, the World Bank, the World Health Organization (WHO), the United Nations Environment Programme (UNEP), the United Nations Development Programme (UNDP), the International Energy Agency (IEA) and the Organisation for Economic Co-operation and Development (OECD).

The commitment has also been made between the scientists of the CM, IAM and IAV communities. The latter originated, broadly speaking, in the initial three-group structure of the IPCC. These communities are more clearly defined, however, by their role in modelling than by their scientific disciplines (physics and dynamics of the climate for the first, socio-economics and biochemistry for the last two). Some of these communities have taken institutional forms that go beyond the IPCC groups and aim to coordinate research outside that framework: this is the case, for example, for the Integrated Assessment Modeling Consortium, set up in 2006 and regrouping 37 teams, or for the Earth System Science Partnership. New organizations for coordinating climate research programmes are also appearing, such as the World Climate Research Programme or the International Geosphere-Biosphere Programme. One of the purposes of these institutions, apart from the targeted objectives that define them, is to share tools: calculating resources and above all databases. This is a technical necessity. It is also a political necessity, because one of the explicit objectives of the reform is to encourage capacities of modelling and analysis in developing countries. This strategy starts with the sharing of tools. The fourth section of the TNS report describes a number of initiatives (meetings, workshops and so on) with the objective of developing collaborative approaches to these subjects (scenarios library, data management, downscaling, integration of mitigation into IAV models, sensitivity to stakeholders' concerns and so on).

The TNS report took particular care in defining the time line of the work of these communities throughout the six years between the fourth report (2007) and the future fifth report, the main objective of the reform being to resolve the contradiction between a pressing demand for fresh studies and the fact that the research and modelling time cannot be compressed. The parallel structure provides the solution to a possible compromise that does not sacrifice either of these needs. The coordination between research teams is not only methodological, but has also become programmatic and managerial (Moss et al., 2008, p. 20).

Far from being abandoned, the category of scenario is extended and used for the diffusion of information to the different levels of the assessment process. By discarding its linear structure, the IPCC has chosen to make the scenarios a shared tool, constructed jointly by the modelling and assessment communities. The scenario is not an alternative to modelling, but its framework of possibility. The concept of scenario is no longer defined by the way it is constructed, or by its structure: the list of principal driving forces, the evolution of the social tensions involved, the hypotheses chosen for a small number of variables. The scenario is defined by what it produces – a certain evolving state of the world and more particularly of GHG concentrations, represented by the RCP – and by what it allows, namely a coordination of the knowledge processes implemented by the different communities. Thus, the TNS report gives no instructions about the method of construction of these scenarios. It prefers to define a framework based on outputs, giving scenarios the objective of shedding more light on a large number of uncertainties, and therefore of intervening at different levels in the cognitive process. This explains the multitude of scenarios that have resulted: benchmark scenarios, stabilization scenarios, technological scenarios, climate scenarios, regional scenarios and so on.

The parallel process generates five different 'products' that almost all have a mixed 'scenario + model' form. The RCPs are described as benchmark scenarios of GHG concentrations. Medium- and long-term climate models are based on the RCP forcing scenarios. The IAM and IAV communities are expected to develop, fairly freely, a wide range of scenarios exploring all the social alternatives in terms of technologies, energy choices and policies of stabilization or adaptation. The same communities are given the task of establishing story lines – stories linking together social evolutions, emissions and impacts and serving as a framework for the responses of policy makers at the global or regional level. The combination of 'physical' climate models and IAM–IAV models produces integrated scenarios that are then available to the decision makers. The sequence begins with the RCPs, followed by the parallel phases – RCP-based climate model ensembles and pattern scaling, new IAM scenarios, and global narrative storylines – leading to the integrated scenarios (Moss et al., 2008, p. 18).

The Epistemology of Climate Risk Scenarios and Models

A paradigm, in the sense of Thomas Kuhn, associates a structured community and a set of cognitive schemes and methods defining the approach favoured by the community. In the case of climate change, we have seen the role that scenarios play in providing an interface between communities using different knowledge and methodologies.

Does the scenario also represent a set of well-defined cognitive schemes? Probably not yet. The epistemology of climate scenarios is far from constituting a completely stabilized corpus of doctrines and far from answering all the questions raised by the successive working groups. I shall therefore draw up a brief list of the questions that remain open, exploring the last of these questions in more detail.

- The disconnection between economic studies and climate studies remains ambiguous. There is a problematic confusion in Box 3.1

proposed between economic scenarios describing the evolution of the economic system and the emissions scenarios that they are partly responsible for producing. They are placed together as IAMs. Although they are obviously connected, they are not merged. The wide variety of economic scenarios can converge on a small number of emissions scenarios. Conversely, these emissions are not exclusively anthropogenic. The timid break of the IAM/CM parallelism in step two of the parallel approach is enigmatic. Jean-Charles Hourcade et al. (2008) also refer to this difficulty in the following terms: 'It would be increasingly blurring to disconnect the emissions scenarios from those used for analysing how to orient the world economy towards one of them. But policy analysis requires much more scenario variants than climate modelling itself' (p. 115).

- It is unlikely that two intermediate RCPs are sufficient for understanding the interactions and feedbacks between all the factors that affect the climate. This is all the more so since these two intermediate RCPs are scenarios of very moderate emissions reductions. It should also be recalled that the first climate simulations made on the basis of emissions rather than concentrations[9] showed that in a hot climate, the oceans and plant cover absorb CO_2 less well (Guillemot, 2007). By basing their work on the RCPs, climatologists run the risk of losing these kinds of feedback effects of the climate on plant cover and the carbon cycle. To what extent can we interpolate the climate changes between different RCPs, that is, between levels of radiative forcing, of which there are now so few? We have seen that interpolation is made very difficult by the non-linear nature of the relationship between forcing and climate change.

- To what extent can a concentration profile be abstracted from the different scenarios of socio-economic evolution? To begin with, this could be done by referring solely to forcing levels. But how will it be possible to associate each of the new scenarios, which will be freely developed by the IAM community, with one sole RCP? The TNS report (p. 22) states: 'Most new IAM scenarios will not have any relationship to the RCPs, given that an RCP is only one scenario created by a simple modelling team. However, some of the new IAM scenarios may be developed to approximate the concentration pathway of an RCP'. There is a sort of '1 to n' relation between RCPs and socio-economic scenarios, the status of which is ambiguous. Will the desire to bring emissions scenarios in line with RCPs constitute a simple possibility, an approximation to be made a posteriori, or a restrictive framework limiting the imagination of the scenario writers?

- One argument put forward in favour of the parallel structure is that climatologists could rapidly make use of the new IAM scenarios without repeating all their simulations thanks to the technique of 'pattern scaling'. This is presented as a means of approximating the profile of a regional climate for a given forcing scenario by using an existing climate simulation.[10] This innovation is therefore considered to enable climate models to be associated a posteriori with new IAM scenarios. It is founded, however, on strong hypotheses that do not allow it to be applied generally.

- How can climate change be treated at the local level? What method should be used for downscaling the results of atmosphere-ocean general circulation models to a regional level such as Europe? Or to the level of large countries like China, India or Brazil? Or to each region of a country like France? The fourth report proposes several statistical and non-statistical methods, but they raise a number of problems. In a complementary way, how can the IAM community supply economic emissions scenarios at a regional level that could then be re-injected into the mesh of climate models? Who will deal with the question of plant cover, to which approaches differ not only between the CM and IAM communities but also within each of them?

- One sole RCP profile of strong reduction to a level of $3W/m^2$, that is, about 490 ppm, is very little if we want to focus on the medium term, understand the feedback effects between global warming and environmental damage, with possible tipping points, non-linearities and long-term effects of short-term policies. As Hourcade et al. (2008, p. 116) wrote: 'The main challenge is the question of the *path dependencies* of development patterns and emissions trends in particular those created by the building and transportation infrastructures in developing countries, investments in the electrical sector, the orientation of R&D, the dynamics of land uses'.

- The report justifies the use of different models for different RCPs. Nevertheless, is it not dangerous, conversely, to define each RCP based on one sole model? Here, one source of uncertainty is drastically but artificially reduced by greatly increasing the model risk, that is, the strong dependence of the results on the specific hypotheses and parameter setting of the model.

- To improve the understanding of uncertainties, can we go beyond this framing between high and low forcing levels? Should we introduce probabilistic models? Can we attribute probabilities to the scenarios? I would like to dwell a little longer on this question, which, as a historian of probability calculus, I find particularly stimulating.

From Uncertainty to Risk Management

In the *Special Report on Emissions Scenarios*, published in 2000, the scenarios are presented as alternative pictures of the future to be used as tools for grasping the numerous *uncertainties* running through them, and to reduce these uncertainties through the qualitative and quantitative study of the effects of driving forces on the climate. These numerous alternative scenarios cannot, however, in any way be considered as elements in a finite set of possible scenarios. It is therefore impossible to ascribe to any given scenario a probability of occurring in a delimited future. Just as there cannot be a 'central', 'average' or 'mean' scenario, so it is inconceivable to speak of the most 'likely' or 'probable' scenario. As the 'Summary for Policymakers' states: 'Probabilities or likelihood are not assigned to individual SRES scenarios' (p. 11). The distribution of scenarios in the cone of possibilities indicated nothing more than a healthy opening-up of conceivable futures, a sort of biodiversity of the imagination that could not be reduced to a probability distribution.

The fourth report introduces a remarkable change in this approach. If the whole set of scenarios cannot be attributed probabilities, the authors start to say that the results of simulations, projections within a scenario, can be: 'Probabilistic characterizations of future socioeconomic and climate conditions are increasingly becoming available, and probabilities of exceeding predefined thresholds of impact have been more widely estimated' (IPCC, 2007a, Ch. 2, p. 135). The three IPCC working groups have sufficiently widely accepted and practised this probabilistic characterization of results to have felt the need to codify the probabilities of future climate events by standardized expressions:[11]

Description of confidence

Terminology	*Degree of confidence in being correct*
Very high confidence	At least 9 out of 10 chance of being correct
High confidence	About 8 out of 10 chance
Medium confidence	About 5 out of 10 chance
Low confidence	About 2 out of 10 chance
Very low confidence	Less than a 1 out of 10 chance

Description of likelihood

Terminology	*Likelihood of the occurrence/outcome*
Virtually certain	> 99% probability of occurrence
Very likely	90 to 99% probability
Likely	66 to 90% probability
About as likely as not	33 to 66% probability

Unlikely	10 to 33% probability
Very unlikely	1 to 10% probability
Exceptionally unlikely	< 1% probability

Source: IPCC (2007a).

The report does not detail the nature of the probabilities of these events. Are they *epistemic* because they are linked to our (lack of) knowledge about the future, or *ontic* because they are inherent to physical and social processes? Nor does it detail the basis of their evaluation: are they *frequentist* because we are treating past observations as the results of a series of tests repeated under identical conditions, or, even more boldly, because we are treating simulations of the future as 'equiprobable cases'? Or are they *subjective*, because the attribution of a probability is the result of an expert's personal evaluation, which appears to be easier to justify in the context of these scenarios? Can this subjective probability be combined and revised on the strength of observations or simulations, adopting a Bayesian approach? We can see that this is a tricky question, bringing us back to centuries of debate about 'contingent futures' reaching back as far as Aristotle, and about the possible justifications of probabilistic interpretation that run from Blaise Pascal and Jacob Bernoulli through to the contemporary works of Ernest Coumet, Ian Hacking and Lorraine Daston.

Most often, today's authors limit themselves to subjective judgements that enable them to ponder the projected results by subjective probabilities. This leads them (GII: IPCC, 2007a, p. 145) to characterize an implausible future by the attribution of a zero or negligible subjective probability, while a plausible future is characterized either by the impossibility of measuring its likelihood or by a non-negligible ascribed likelihood. This explains the origins of the three categories: implausible futures, plausible futures without ascribed likelihood and plausible futures with ascribed likelihood (Carter et al., 2007, p. 145). The authors classify thought experiments, simulations and scenarios of impossible futures in the first category and all the other techniques in the second. In this, they follow the majority of researchers in the IAV community who maintain that climate change is characterized by deep uncertainty about the initial parameters, the laws of evolution of each component and their interactions, making judgements about probability impossible.

More rarely, we can move into the third category, illustrated by probabilistic projections: a probability is associated with the results of simulations of variants of the same scenario (with the same model), in which a few parameters are varied with the aim of measuring the sensitivity of the system – or rather the model – to those parameters. Using the Monte Carlo method, it is then possible to construct a frequency distribution of certain results, which can then be considered as an estimation of a probability law. Even more rarely, models are written a priori in a probabilistic form. That is to say, they incorporate probabilistic hypotheses about the main exogenous variables or about the random error terms. In this way, probability densities have been defined for certain socio-economic inputs, and probability densities have then been deduced for climate outputs such as global temperature or precipitations.[12]

Chapter 19 of the GII report 'Assessing Key Vulnerabilities and the Risk from Climate Change' goes far in the probabilistic modelling of dangerous anthropogenic interference with the climate system, which is the expression of the UNFCC. In particular, it proposes a synthesis of the levels of confidence in different values of global warming based on several studies used for the conclusions of the 'Summary for Policymakers'. These conclusions are, for example, of the type: 'it is very likely (more than 90% probability) that global warming exceeds 1.5°C and likely (66% to 90% probability) that it will remain within the interval 2°C–4.5°C'. Or, more complicated: global warming of 2°C above pre-industrial levels (1.4°C since 1990) is unlikely (probability < 33%) if the concentration of CO_2 is lower than 350 ppm for all the scenarios, or less than 410 ppm for 50 per cent of the scenarios. This is roughly what we can read from the curves (Schneider et al., 2007, p. 801). We can see, however, that these conclusions combine the probabilistic projection of each modelling with the 'model risk'. Moreover, the authors stress the fact that these conclusions remain highly dependent on the shape of the distribution, especially the tail above 4.5°C warming, which remains relatively unknown, and the normal approximation of which is no doubt inaccurate: 'we assign no more than low confidence to any of the distributions or results presented in this section, particularly if the result depends on the tails of the probability distribution for climate sensitivity. Nevertheless . . . a risk-management framework requires input of (even if low-probability, low-confidence) outlier information' (p. 801).

Despite these reservations, if an effort has been made over the last few years to shift imperceptibly from the uncertainty that has caused so much concern to risk – a combination of a set of values of gains and losses and a measurement of probability – this is because reasoning in terms of risk offers more than just epistemological advantages. Above all, it offers important *political* advantages.

- It allows a series of practices by which risk can be tamed, including calculation. First, Pascal, who in the seventeenth century claimed to have invented the 'geometry of chance', pioneered the calculation of an *expectation*. Second, probability allows for the calculation of a *variability* or volatility of the cost of impacts, which climate studies have yet to make much use of. Third, probabilization could be used to put a figure on *inequalities* between different populations in terms of vulnerability. But that is yet to come. Today, researchers limit themselves to expected values and cautious allusions to distributions, without yet daring to draw the conclusions.

- Risk is expressed in monetary terms, making it possible to perform economic cost–benefit calculations under uncertainty that both economists and policy makers find convenient.
- This type of reasoning suits politicians, who know that risk-taking is part of their everyday life as decision makers.
- It suits firms, which have made risk management a branch of their activity.
- It brings to light extreme risks of high impact and low probability.

Among the new methods for assessing impacts and vulnerabilities, alongside scenarios and IAMs, the fourth IPCC proposes cost–benefit analyses, sensitivity analyses, spatial analogies, indicator approaches, economic modelling, technological forecasting and . . . risk management. So in the space of 10 years, we have moved from the battle over reducing uncertainties to the assessment of climate risk, and then from the assessment of those risks to their management. The seed had been planted as early as the 1980s when the integrated assessment methods discussed in the forums of IIASA included optimized risk management by the communities concerned. This last dimension was still posing problems in 2007, however. In the scientific arena, many authors still maintain that a scientific treatment of uncertainty is necessary, rather than a decision-making approach that seeks simply to manage the uncertainty by prioritizing goals and actions.

The *risk assessment* highlighted by the fourth report (GIII: IPCC, 2007b) is seen as a new approach, widening the range of existing assessment methods. The report quickly moves, however, from risk assessment to *risk management*, defined as 'the culture, processes and structures directed towards realising potential opportunities, whilst managing adverse effects'. In other words, it is seen as a system of direct response to the potential dangers of the climate, implemented by the decision makers. In practical terms, the two responses proposed are the two categories of mitigation and of adaptation to climate change. This decision-making approach has the advantage of speaking directly the language of the decision makers and providing them, for each scenario, not only with the pair of probabilities and damages that precisely define a risk, but also with the range of political and organizational responses with which it can be met.

For example, as Carter et al. suggest, future states of global warming can be directly associated with the zones of maximum benefit of adaptation and mitigation, and more broadly with the optimal strategies adapted to the scale of the risks (Carter et al., 2007, p. 140).

Is there an advantage for scientists in speaking the language of managers in this way? Nothing could be less certain. The receptiveness of the decision makers is not necessarily related to a better understanding of the uncertainties, and there is often an increased risk of over-interpretation: reading more into the studies than what they can actually say is another new risk.

AN UNEXPECTED VISITOR. NINE ARGUMENTS FOR TAKING THE ECONOMIC CRISIS INTO ACCOUNT

The IPCC Fourth Assessment Report was published before the sub-prime crisis that began in the spring of 2007 and suddenly turned into a worldwide financial crash in September 2009 after the collapse of the US mortgage finance lenders Fannie Mae and Freddie Mac in July and of the banks Northern Rock (saved by nationalization) and above all Lehman Brothers (abandoned by the public authorities and declared bankrupt on 15 September). This financial crisis, the macro- and microeconomic origins of which lie in the extreme indebtedness of the US economy, has not run its course, and its medium-term effects cannot yet be calculated. We already know nonetheless that it has triggered a global recession.[13]

It appears that the different climate science communities – used to reasoning on the premises of 'business as usual' or shocks driven by active long-term policies but not on endogenous shocks of the economic system – have not yet taken the full measure of the implications of such a brutal disruption of the global economic environment for their prospective studies and for policies of mitigation and adaptation. The impacts of this economic and financial crisis are of such a scale, however, that it will be difficult to continue reasoning in the same way.

At present, we lack perspective. If we look at the reactions of the daily and general-interest press to the links between financial crisis and environmental crisis, or the conclusions of the Conference of the Parties in Poznan in December 2008, we can already observe the terms of a first controversy about the environmental opportunities of the crisis. The first common-sense arguments are of a pessimistic nature.

Argument 1 The financial crisis has pushed the ecological crisis and climate change into the background. It has sparked a global economic crisis that will monopolize concerns and resources for some time to come. The consumer and credit crisis is ongoing. The fall in consumption is leading economies into recession. Growth is under threat, and there is therefore little reason to worry about limiting it.

Argument 2 The crisis has forced governments to spend substantial sums of money to recapitalize banks. The priority has now shifted to supporting the industrial activity of sectors in crisis, for example, the car industry. As governments have emptied their coffers, there is nothing left for budgets of mitigation or adaptation to climate change.

Argument 3 Because of the crisis, countries in difficulty, such as Italy, Poland and Germany, are asking for a suspension of the Kyoto process rather than its revival. Developing countries are angry about the false declarations of the industrialized countries: their coffers were not empty after all! A climate of mistrust is developing, and the chances of bringing the developing countries on board for climate negotiations are dwindling. This was the crucial issue at the Conference of the Parties in Poznan, with the risk of the energy–climate package proposed by the European Union falling by the wayside.

The same crisis that broke during the last quarter of 2008 has nonetheless also inspired very different, much more optimistic observations.

Argument 4 The contraction in economic activity will automatically cause a reduction in GHG emissions. The ideological debate about growth or de-growth has suddenly been overtaken by stagnation, or even recession. What the politicians could not do, the crisis will! Even if it only lasts a few quarters, the drop in consumption and therefore in GHG emissions could mark a turning point and induce more long-lasting behaviour of a prudent and frugal nature.

Argument 5 The reflationary policies that will have to be adopted provide an excellent opportunity to establish a carbon-light economy. Governments must react to the economic crisis either by boosting consumer spending by increasing purchasing power, which they generally try to avoid for fear that the main beneficiaries will be exports from emerging countries, or by a programme of support for investment and the launching of large public works. The investments required for sustainable growth, in the building and transport sectors, for example, could benefit from this sort of reflationary plan. A number of economists, including Nicholas Stern, affirm this:[14] 'we must invest in infrastructure for alternative means of producing electricity, and in energy-efficient buildings, so that we can move to a carbon-light economy. There is an enormous opportunity for such investments, probably about a trillion dollars a year over the next 20 years.' The crisis offers a possibly unique opportunity to respond to the economic and the ecological crises with the same action, and to reconsider the mode of development in Europe or the United States as in the developing world.

Argument 6   We were in a deadlock. The crisis has overturned geopolitical balances and reshuffled the deck because of differential vulnerabilities to the contraction of markets and the recession, of which we do not yet know all the repercussions. Will the United States resume its position as leader? Will the rest of the world continue to finance its debt? Will China emerge relatively unscathed or will it be dragged down too? Will Africa continue to tie its destiny to the new emerging countries? If we accept that it is necessary both to define the common rules and to adapt the three political responses to climate change – mitigation, adaptation and innovation – to suit the specificities and vulnerabilities of each country, then the effects of the crisis will weigh directly on the negotiations. The successor to the Kyoto process, which must be found before the Copenhagen meeting (7–18 December, 2009), will benefit from the position of the new US administration, but it also depends on the new global economic situation. In other words, the climatic and ecological vulnerability of each country, which conditions its commitment to any process of mitigation and adaptation, is highly dependent on its economic vulnerability, which the economic crisis has reconfigured.

A third point of view to be developed concerns not so much the opportunities, but the lessons that we can draw from an analysis of the crisis, its causes, development and governance, and the available responses and levers of action. In more than one respect, there are similarities to be exploited between financial risk and climate risk, beyond the processes themselves, which are very different, as the latter depends on a partly natural system, over a much longer time scale, making it less accessible to regulation.

Argument 7   Is it still possible to take balanced growth as the only reference for economic scenarios and models? Financial markets, raw materials markets, food markets – and perhaps currency markets, one day – have shown that they can enter periods of high volatility (in terms of volumes traded, but above all prices) completely outside the range of any forecasts. Is it possible, therefore, to continue working within a framework of general equilibrium, or even partial equilibrium in each market? Most IAMs are still based on a model of balanced growth with intertemporal choices in the style of Ramsey, or a calculable general equilibrium model. Before this same crisis, Hourcade et al. (2008) stressed the need to move away from optimal baselines and take into account the imperfections and quirks of the real economic machinery, in a world that is more often shaken than still. They called for the development of scenarios with economic disequilibrium generated by the interplay between inertia of social and technical systems, imperfect foresights and 'routine' policy behaviours, in order to detect the many sources of suboptimality (structural debt, unemployment, informal economy and unfulfilled basic needs, capacity shortages, missing markets). We can now add to the list the particular financial rationality revealed by the crisis, with its astronomical payments, speculative bubbles, and the most ordinary financial criminality that appears to be more the rule than the exception.

Argument 8   The financial crisis of autumn 2008 has revealed a number of weaknesses in the management of financial markets, and in particular:

- a lack of market regulation of high-risk products;
- a 'cascade' structure of risk coverage by securitization, producing financial assets that are almost impossible to trace or control, but the excessive risk of which was accepted by the actors with direct and personal interest in the profits, including credit rating and reporting agencies;
- a failure to take into account rare and costly events in the models. In particular, an important model risk stems from incorrect hypotheses about the distribution of risks and the characteristics of their processes. Over several decades now a number of specialists have been challenging the combined hypotheses of Gauss's law, independence and Brownian motion that are used to model stock markets and some derivatives markets.[15] Nevertheless, they are still forming the basis for programmes of portfolio management and option pricing.

It is worth meditating on the ingredients of the financial market crisis and developing analyses to bear on the management of climate risk.

- Taking into account all the climate risks, including the most extreme, is the pre-condition for an effective policy response, as Stern told a journalist from *Le Monde* on 21 October 2008 (see n. 13): 'There is a lesson to be learned in the financial crisis. If we ignore the risks that develop in a system, we end up in serious trouble. This is a powerful lesson for climate change, the consequences of which, if we do not act, will be far greater than the current crisis'.

- We must take into consideration the 'model risks', that is to say the consequences (costs and probability) of an incorrect formulation of hypotheses in the models of projected emissions and of the resulting warming. This calls for the utmost care to be taken in the probabilistic modelling of the uncertainties.

Argument 9   The carbon market is a financial market. It has the potential to regulate transactions and establish a 'fair' price per ton of carbon. It also possesses, however, all the risks of loss of control that we have just experienced with the sophisticated financial products of securitization, and we must make provision against those risks as from today. Numerous studies have already raised the alarm about several dysfunctions.

- Malfunctioning of the Clean Development Mechanism (CDM). This market, worth €10 billion in 2007 and involving slightly more than a thousand projects, has already attracted a number of predators who extract unjustified profit from it. The project owners pay those who validate projects submitted to the CDM Executive Board, recalling the analogous situation of credit ratings agencies having an interest in the profits derived from securitization of subprime mortgages. This cannot fail to raise concerns.
- The respect of emissions quotas under pain of having to buy credits from someone else is a severe rule, much disputed. Will it be accepted? Will it be evaded? What would constitute 'fair' initial quotas? By what mechanism will they be enforced?
- The carbon market is not free from the risk of corruption, as participants at the 13th International Anti-Corruption Conference held in Athens in November 2008 pointed out: 'There are significant and growing risks at all stages in the process'. On this subject, the possible embezzlement of funds devoted to the prevention of deforestation has been mentioned.

CONCLUSION

My analysis in this chapter shows that an important paradigm shift has taken place, both in terms of the coordination of the work of the different scientific communities and in terms of response to the demand of politicians engaged in climate negotiations or other international negotiations related to development and the environment. The analysis also reveals that this new paradigm redefines several types of scenarios from which it will be difficult to form a coherent whole, leaving several areas of shadow in our knowledge of climate change, corresponding to unresolved uncertainties.

 I have focused more closely on the tendency to treat these uncertainties as risks, associated with possible futures to which probabilities can be ascribed, in particular because the new risk management approach that calls for a shift from an epistemic to a managerial perspective neglects three fundamental questions. Risk assessment requires the prior assessment of possible future states of the world, and therefore of the impacts of climate change. The first question that lacks a satisfactory solution is therefore that of the evaluation of the costs of future impacts. This raises the problem of the moral and economic conventions that must be established concerning the burden to be imposed on future generations, and the expression of this arbitrage in the form of an intertemporal utility function and/or a fair discount factor for these costs. All the cost–benefit studies that have sought to answer the political question of 'where and when to intervene' encounter this difficulty, the most striking example being the monumental Stern Review, often debated on the grounds of this unique viewpoint (Stern, 2006). It appears that the current tendency is to reject the idea of calculating any long-term discount factor. The second question, less often discussed, turns on whether or not it is possible to ascribe degrees of likelihood or probabilities to the socio-economic factors of radiative forcing and to the resulting scale of climate change. Here we come up against the difficulties of frequentist, logical or subjective justifications for ascribing probabilities to possible futures that do not result from a simple game of chance. The third question, raised by the shift from risk assessment to risk management, presupposes either that we have solved the previous two problems or that we have bypassed them by managing the uncertainty, which is no longer an issue of calculation but of moral prudence. Here we can find solutions drawn from political philosophy, such as the precautionary principle or 'enlightened catastrophism' (Dupuy, 2002; Godard, 1997), illustrating, in a way, two typical but contrasting attitudes towards the main dangers: irrational activism or equally irrational fatalism. Here again, the concept of scenario appears to provide the narrow path between two precipices: fascination for the worst on the one side, and a utilitarian and conservative vision, too blinkered by the modelized projection of present structures, on the other. As Olivier Godard (2007) has shown, however, there is still some way to go before modellers, decision makers and citizens all share, without ambiguity or diversion, the culture of the scenario and more particularly its pluralist vision.

 The most serious warning, however, is addressed more directly to economists and conflicts with what often bonds together their community:

the belief in an economic system in equilibrium solely on account of the hypothesis of the rational choice of agents. The economic and financial crisis that is currently shaking the world brings us daily proof that the financial market – supposed to be the most fluid and the one that comes closest to the concept of a perfectly competitive market – is in fact the object of actions, none the less rational, by agents who, by maximizing their immediate profits, wreck the possibilities of equilibrium and of optimizing the general interest. Flagrant speculation, distribution of unjustified incomes, artificial tax havens, trafficking in 'toxic' products, conflicts of interest among controllers – everything points to the fact that the world of finance is not in equilibrium, but in a state of permanent disequilibrium, and what is more, it is moving ever further away from satisfying the general interest. It is therefore important to draw all the relevant conclusions in terms of economic calculation and assessment. Models must take into account the extreme diversity of agents' ways of reasoning, the reality of mechanisms in every sector, including the shadow economy, the two faces – beneficial and destructive – of all innovation, and the non-negligible possibility of major risks. Failing these, our scenarios will only be of illusory wisdom or audacity, because we are certainly not experiencing balanced, equitable and well-regulated development.

NOTES

1. Ramsey's optimal growth models, Koopmans's activity analysis models, Forrester's dynamic and systemic models for the Club of Rome.
2. For example (IPCC, 2007b, p. 174), 'The SRES report (Nakicenovic et al., 2000) defines a scenario as a plausible description of how the future might develop, based on a coherent and internally consistent set of assumptions ('scenario logic') about the key relationships and driving forces'.
3. This is confirmed by the exchanges between climatologists and economists observed during seminars held at the Alexandre Koyré Centre, Paris, in which I participated.
4. There remains the difficulty of comparing evaluations made in different currencies: the exchange rate and the purchasing power parity methods remain divergent, and the latter also produces divergent evaluations depending on whether one refers to the OECD, Eurostat or the World Bank.
5. See http://www.mnp.nl/ipcc/pages_media/meeting_report_workshop_new_emission_scenarios.pdf.
6. World Climate Research Programme (WCRP); International Geosphere-Biosphere Programme (IGBP); Task Group on New Emission Scenarios (TGNES); Task Group on Scenarios for Climate and Impact Assessment (TGCIA); Analysis, Integration and Modeling of the Earth System (AIMES).
7. See http://www.mnp.nl/ipcc/docs/index0407/Backgroundpaper_2007Sept11_final.pdf.
8. Parameterization consists in replacing physical phenomena that are poorly understood or on a scale very much below that of the general-circulation-model mesh, such as convection and condensation linked to the cloud system, by sets of parameters on the scale of the mesh (see Guillemot, 2007).

9. These first simulations were made by the Institut Pierre Simon Laplace des Sciences de l'Environnement (IPSL) and the Hadley Centre in 2001.
10. 'Pattern scaling assumes that within limits, the regional pattern of change in some variables (for example, temperatures) can be tuned to correspond with a higher or lower level of forcing than the one used in the original simulation. Thus, if simple CMs can be used to define the global mean temperature response to a given radiative forcing, the pattern of climate change produced by AOGCMs or ESMs for the RCP giving the closest radiative forcing to the target can be scaled linearly upward or downward according to the ratio of the simulated global mean temperature change for the RCP and the temperature change defined in the simple CM for the target radiative forcing' (Moss et al., 2008, p. 25).
11. 'Virtually certain >99%, extremely likely >95%, very likely >90%, likely >66%, more likely than not >50%' (IPCC, 2007a, Ch. 2, Box 2.4).
12. See, for example, Den Elzen and Meinshausen (2006), who used various densities of probabilities in their simplified MAGICC climate model to estimate a relation between the probability of reaching certain climate target values and certain emissions reduction levels.
13. On the financial and economic crisis, see MacKenzie et al. (2007), Walter and Brian (2007), Artus et al. (2008), Sapir (2008), Lordon (2008) and Armatte (2009).
14. Nicholas Stern, quoted by Hervé Kempf, 'L'environnement pourrait tirer bénéfice de la crise économique', *Le Monde*, 21 October 2008; our translation.
15. See in particular the French school – from the mathematician Benoit Mandelbrot (1961, 1997) to Daniel Zajdenweber (1976) and Christian Walter (1994; Walter and Brian, 2007) – which has developed the alternative paradigm of 'wild randomness', fractal time in stock markets, and scaling, alpha-stable Pareto–Levy laws with fat tails and infinite variance.

REFERENCES

Alcamo, J., D. van Vuuren, C. Ringler, J. Alder, E. Bennett, D. Lodge, T. Masui, T. Morita, M. Rosegrant, O. Sala, K. Schulze and M. Zurek (2005), 'Methodology for developing the MA scenarios', in S.R. Carpenter, P.L. Pingali, E.M. Bennett and M.B. Zurek (eds), *Ecosystems and Human Well-Being: Scenarios. Findings of the Scenarios Working Group*, Washington, DC: Island Press, Chapter 6, pp. 145–72.
Armatte, M. (2005), 'Economical models of climate change: the costs and advantages of integration', in P. Freguglia (ed.), *The Sciences of Complexity: Chimera or Reality?*, Bologna: Ed. Esculapuio, pp. 51–71.
Armatte, M. (2007), 'Les économistes face au long terme: l'ascension de la notion de scénario', in A. Dahan Dalmedico (ed.), *Les Modèles du futur*, Paris: La Découverte, pp. 63–90.
Armatte, M. (2008), 'Climate change: scenarios and integrated modelling', *Interdisciplinary Science Reviews*, 33 (1), pp. 37–50.
Armatte, M. (2009), 'Crise financière: modèles du risque et risque de modèle', *Mouvements*, No. 58, 2009-2, pp.160–76, available at: http://www.mouvements. info/Crise-financiere-modeles-du-risque.html, accessed 17 June 2009.
Armatte, M. and A. Dahan Dalmedico (2004), 'Modèles et modélisations (1950–2000): nouvelles pratiques, nouveaux enjeux', *Revue d'Histoire des Sciences*, 57 (2), pp. 245–305.
Artus, P., J.-P. Betbèze, C. de Boissieu and G. Capelle-Blancard (2008), *La*

crise des subprimes, Rapport du Conseil d'Analyse Economique, Paris: La Documentation Française.

Barker T., et al. (2007), 'Technical summary', in IPCC, *Climate Change 2007: Mitigation. Contribution of Working Group III to the Fourth Assessment Report of the Intergovernmental Panel on Climate Change*, edited by B. Metz, O.R. Davidson, P.R. Bosch, R. Dave, L.A. Meyer, Cambridge, UK and New York: Cambridge University Press.

Carter, T.R., R.N. Jones, X. Lu, S. Bhadwal, C. Conde, L.O. Mearns, B.C. O'Neill, M.D.A. Rounsevell and M.B. Zurek (2007), 'New assessment methods and the characterisation of future conditions', in IPCC, *Climate Change 2007: Impacts, Adaptation and Vulnerability. Contribution of Working Group II to the Fourth Assessment Report of the Intergovernmental Panel on Climate Change*, edited by M.L. Parry, O.F. Canziani, J.P. Palutikof, P.J. van der Linden and C.E. Hanson, Cambridge, UK: Cambridge University Press, pp. 133–71.

CPB (1992), *Scanning the Future: A Long-Term Scenario Study of the World Economy 1990–2015*, The Hague: SDU Publishers.

Dahan Dalmedico, A. (2007), 'Le régime climatique, entre science, expertise et politique', in A. Dahan (ed.), *Les Modèles du futur*, Paris: La Découverte, pp. 113–39.

DATAR (1971), 'Scenario de l'inacceptable. Une image de la France en l'an 2000', *Travaux et Recherches de Prospective*, No. 20.

Den Elzen, M. and M. Meinhausen (2006), 'Meeting the EU 2°C climate target: global and regional emission implications', *Climate Policy*, **6** (5), pp. 545–64.

Dupuy, J.-P. (2002), *Pour un catastrophisme éclairé. Quand l'impossible est certain*, Paris: Le Seuil.

Godard, O. (ed.) (1997), *Le principe de précaution dans la conduite des affaires humaines*, Paris: MSH-INRA.

Godard, O. (2007), 'Pour une morale de la modélisation économique des enjeux climatiques en contexte d'expertise', in A. Dahan Dalmedico (ed.), *Les modèles du futur*, Paris: La Découverte, pp. 203–26.

Guillemot, H. (2007), 'Les modèles numériques du climat', in A. Dahan Dalmedico (ed.), *Les modèles du futur*, Paris: La Découverte, pp. 93–112.

Hare, B. and M. Meinshausen (2005), 'How much warming are we committed to and how much can be avoided?', *Climatic Change*, **75**, pp. 111–49.

Hourcade, J.Ch., Z. La Rovere, P.R. Shukla and T. Kejun (2008), 'Proposal for the next Vintage of Long Run Scenarios in a Changing Scientific and Policy Context', in R. Moss et al., *Towards New Scenarios for Analysis of Emissions, Climate Change, Impacts and Response Strategies*, Geneva: IPCC, pp. 115–24.

IPCC (2007a), *Climate Change 2007: Impacts, Adaptation and Vulnerability. Contribution of Working Group II to the Fourth Assessment Report of the Intergovernmental Panel on Climate Change*, edited by M.L. Parry, O.F. Canziani, J.P. Palutikof, P.J. van der Linden and C.E. Hanson, Cambridge, UK: Cambridge University Press.

IPCC (2007b), *Climate Change 2007: Mitigation. Contribution of Working Group III to the Fourth Assessment Report of the Intergovernmental Panel on Climate Change*, edited by B. Metz, O.R. Davidson, P.R. Bosch, R. Dave and L.A. Meyer, Cambridge, UK and New York: Cambridge University Press.

Jones, R.N. (2004), 'Managing climate change risks', in J. Corfee-Morlot and S. Agrawala (eds), *The Benefits of Climate Policies: Analytical and Framework Issues*, Paris: OECD, pp. 251–97.

Lordon, F. (2008), *Jusqu'à quand? Pour en finir avec les crises financières*, Paris: Raisons d'agir.
Mackenzie, D., F. Muniesa and L. Siu (eds) (2007), *Do Economists Make Markets?*, Princeton, NJ: Princeton University Press.
Mandelbrot, B. (1961), 'Stable Paretian random functions and the multiplicative variation of income', *Econometrica*, **29**, pp. 517–43.
Mandelbrot, B. (1997), *Fractales, hasard et finance*, Paris: Flammarion (Coll. Champs).
Moss, R. et al. (2007), 'Towards new scenarios. Background for participants', available at: http://www.mnp.nl/ipcc/docs/index0407/Backgroundpaper_2007Sept11_final.pdf, accessed 10 December 2008.
Moss, R. et al. (2008), *Towards New Scenarios for Analysis of Emissions, Climate Change, Impacts, and Response Strategies*, Geneva: Intergovernmental Panel on Climate Change.
Nakicenovic, N. and R. Swart (eds) (2000), *Special Report on Emissions Scenarios. A Special Report of Working Group III of the Intergovernmental Panel on Climate Change*, Cambridge: Cambridge University Press.
OECD (1979), *Interfutures: Facing the Future, Mastering the Probable, Managing the Unpredictable*, Paris: OECD.
Raskin, P. (2005), 'Global scenarios: background review for the Millennium Ecosystem Assessment', *Ecosystems*, **8** (2), pp. 133–42.
Sapir, J. (2008), 'Une décade prodigieuse. La crise fianciére entre temps court et temps long', *Revue de la régulation*, 2nd semester, available at: http://regulation.revues.org/document4032.html, accessed 25 May 2009.
Schneider, S.H., S. Semenov, A. Patwardhan, I. Burton, C.H.D. Magadza, M. Oppenheimer, A.B. Pittock, A. Rahman, J.B. Smith, A. Suarez and F. Yamin (2007), 'Assessing key vulnerabilities and the risk from climate change', in IPCC, *Climate Change 2007: Impacts, Adaptation and Vulnerability. Contribution of Working Group II to the Fourth Assessment Report of the Intergovernmental Panel on Climate Change*, edited by M.L. Parry, O.F. Canziani, J.P. Palutikof, P.J. van der Linden and C.E. Hanson, Cambridge, UK: Cambridge University Press, pp. 779–810.
Stern, N. (2006), *The Economics of Climate Change*, available at: http://www.hm-treasury.gov.uk/sternreview_index.htm, accessed 1 June 2009.
Walter, C. (1994), *Les Structures du hasard en économie: efficience des marchés, lois stables et processus fractals*, Paris: IEP.
Walter, C. and E. Brian (eds) (2007), *Critique de la valeur fondamentale*, Paris: Springer.
Zajdenweber, D. (1976), *Hasard et Prévision*, Paris: Economica.

4.   In defence of sensible economics[1]

Thomas Sterner

UNCERTAINTY IN CLIMATE AND IN THE STERN REPORT

Many people[2] are critical of the role of conventional economics in integrated assessments of large-scale, risky issues such as climate change. While I share many of the concerns in general, I would lean somewhat in the other direction. As suggested by the title of this chapter, I believe that economic models can help clarify some of the very difficult ethical and philosophical issues involved, at least if done sensibly.

One key phenomenon that has been the subject of much discussion in this area is the degree of uncertainty and the handling of this uncertainty. It should nevertheless also be said at the outset that some things are fairly certain. There is, for example, reasonable certainty that an anthropogenic increase in the natural level of radiative forcing is caused by the total quantity of greenhouse gases emitted, which is gradually leading to an increase in average global temperatures. The rate of warming that corresponds to any given increase in greenhouse gases is, however, uncertain, and the biological and other effects of this warming even more so. The highest degree of uncertainty concerns the economic costs of this human-induced climate change.

In the Stern Review (Stern, 2006) – and even more so in the debate that the report has sparked – there is a very wide range of costs of climate change. Some of this uncertainty is due to biogeophysical uncertainties. These uncertainties are built in part on physical feedback effects, such as, for instance, the possible release of vast amounts of methane that are currently trapped in permafrost. As climate change thaws the permafrost, it may lead to the release of the methane, which would positively reinforce the warming itself, resulting in a kind of vicious circle. Similar mechanisms and uncertainties pertain to the changing albedo of Earth, particularly as ice caps melt, and changes occur in cloud formation, biological feedback mechanisms, and so forth.

In the face of all these natural science uncertainties, it is striking that one of the biggest sources of numerical variation in this kind of integrated assessment actually comes from the discount rate. The reason for this is, of course, that virtually all the costs being discussed will occur far into the future, and thus they must be discounted.

DISCOUNTING, BUT ALSO RELATIVE PRICES

Discounting builds on the simple fact that money earns interest. In a growing economy, an investment of say 100 units will in general be expected to give a return such that the accumulated value of the capital invested plus dividends grows exponentially. With a real interest rate of 6 per cent (assuming zero inflation here), the capital will have increased to 106 units after one year. For this reason, we say that a cost of 106 units in one year's time is equivalent to 100 units today.

The rest is arithmetic – but powerful none the less. Capital will double within 12 years and multiply more than 300 times in a century. Discounting in its simplest form can be thought of as the inverse of this growth. Thus, the value of a cost of US$1 billion in 500 years with 6 per cent discounting would be 0.02 cents today. Not only is this clearly insignificant, but there are alarming details, such as the apparent non-linearity of the effects. If I had used a 5 per cent discount instead, the value of the US$1 billion would have been 2 cents instead. It is still insignificant, but it is important to note that the difference between a 5 and a 6 per cent discount rate can change the present value, as in this case, by a factor of 100. This is much greater than the range of variation considered for the climate sensitivity parameter, and discount rates can vary much more than that – maybe from 1 to 10 per cent – so this is clearly an area where we need to be particularly careful.

Discounting appears to suppress the concerns raised by environmentalists involving future environmental damages caused by activities today. For some people, this motivates a rejection of the concept of discounting or of cost–benefit analysis in general. In my view, this is throwing out the baby with the bathwater. Instead we should think carefully about two issues. First, is growth possible for many hundreds of years? Such growth is the fundamental mechanism behind discounting, and hence a necessary requirement for discounting. Second – a somewhat related thought as I shall show later – are there any counteracting forces? One such possibility is, of course, changes in prices over time. Suppose we are interested in a correct economic valuation today of a future cost of V at a time t years hence. If we agree on the principle of discounting, we might still argue that the costs will follow the formula:

$$V_t = V_0(1 + r)^{-t}(1 + p)^t. \tag{4.1}$$

In this formula, we see clearly the possibility that the effect of discounting could be counteracted by changes in relative price if we had reason to believe that the object in question would be appreciating rapidly in value.

The effect of relative prices could be as big as that of discounting! I shall now turn to the issue of when and if this may be a reasonable assumption. I shall start with a general speculative discussion before coming to a more formal argument.

WHAT WILL THE FUTURE LOOK LIKE?

Ultimately, discounting hinges on what view we have of the distant future. In projects that concern only a few years, these issues do not necessarily surface with such intensity. When we speak of problems that will unfold over centuries, however, we need to go back to first principles. With 3 per cent growth, we would be twice as rich in 24 years and almost 20 times as rich in a century. We need to reflect on what this would mean? Would we really be consuming 20 times as much of all the goods and services? Do we today consume 20 times as much of all goods and services as we did a hundred years ago? It is immediately clear that there are some exceptions, such as food. We do not eat 20 times as much food (although our consumption of meat has increased, which does mean that the indirect amount of acreage appropriated – through pastures and so on – has increased more than we might think). I shall argue that there will be many fundamental changes to our consumption basket.

WHEN WILL EVERYONE HAVE A MAID?

Let us take the example of domestic labour and other personal assistants, maids, servants, errand boys and private secretaries. A hundred years ago, I estimate that some 5 per cent of the population in European metropolises, such as London or Paris, had a maid and often a number of other personal employees, such as gardeners, secretaries, chauffeurs and so on (I shall use the word 'maids' as shorthand for all of the above). Since then, incomes have been rising rapidly, and nowadays the middle classes comprising up to maybe half the population have in many ways attained incomes and living standards that only the richest had around the turn of the last century (1900). 'Even' for the working class, factory workers, carpenters or nurses of today, it is not exceptional to be able to afford vacations abroad, cars or houses. I do not intend to make any value judgement by the quotation marks around the word 'even'; they are there to remind the reader just how unthinkable this big increase in consumption and living standard was in the nineteenth century – and still is in low-income countries today. So consumption has soared, yet the number of people who have maids has probably decreased significantly. All this is well known and appears trivial, but I want to dwell on the mechanism so that we can learn from it for the future.

There may in fact be several factors at play. The idea of having maids may be 'out of date'. There may be changes in preferences that are related to historical and social circumstances. As I understand it, however, the main mechanism – why the 'consumption' of maids does not go up when income goes up – is simply one of relative price: the cost of maids is going up at least as fast as incomes are! The example is chosen to be clear and pedagogical, because of course the 'price' to the buyer is almost exactly equivalent to the 'income' of the maid. Assuming that the income distribution is unchanged, the price of (domestic) labour would thus be rising exactly at the speed of the general increase in incomes. If anything, the past century saw some decrease in inequality. This (together with other factors, such as tax wedges created by an increasing public sector, social security and so on) probably implied that the price of maids increased faster than average incomes, explaining a decline rather than an increase in the extent of domestic services.[3]

THE VALUE OF FOOD

A further illustration of the importance of relative prices is afforded by considering the value of food. According to the statistics cited in the Stern Review, the share of the agricultural sector in GDP for the world as a whole is about 24 per cent. This figure builds, of course, on marginal cost pricing, and we can use it to calculate the value of a very small increase or decrease in the sector. Suppose we lose 1 per cent of the produce next year. The loss could very approximately be calculated as 1 per cent of 24 per cent, that is 0.024 per cent of world GDP.

Now assume that we lose 95 per cent of world agriculture. Using the same method of calculation, we would get 0.95*0.24 = 23 per cent of world GDP. This is just about as much nonsense as the idea that everyone in a society could have a maid, and for the same reason: we have again neglected the role of relative prices. A situation in which we really did lose 95 per cent of world agriculture would be an unprecedented disaster leading to widespread strife and suffering and most likely war. A cool-headed economist would say that the price of the remaining food would rise so fast that this – now very small physical amount of food – would be worth more than all the food before it and would in fact come to occupy virtually the whole of world 'GDP' in that situation.

FUTURE ECOSYSTEM SCARCITIES

As we look into the future, we have little exact knowledge. We do know, however, from the Intergovernmental Panel on Climate Change (IPCC, 2007) that a considerable number of ecosystems are threatened to a degree that depends largely on the extent of radiative forcing. One of the more obvious or direct consequences is the change in temperature, which in turn will cause changes in precipitation, the melting of snow and ice, and alterations in storm patterns.

There is already a visible trend towards winters without snow, which will reduce the opportunity for skiing for those who enjoy – or live off – this and other winter sports. The industry strives to survive by making 'artificial' snow by freezing water from lakes or other sources. This is done on a large scale – it is reportedly one of the biggest cost items for some ski resorts in Sweden – using much more electricity than the lifts, and incurring costs that approach a third of the income of some resorts. Even as far north as Piteå – on the Arctic Circle – artificial snow is made to keep the skiers happy! More seriously, it is also virtually certain that a majority of coral reefs will disappear. This is not only a tragedy for beauty, recreation and tourism, but also for storm protection, biodiversity and fish production. Furthermore, the changes in climate are likely to negatively affect water availability in Africa through increased droughts – making agriculture virtually impossible in large parts of the continent – and in Asia through the disappearance of glaciers that regulate stream flow in the major rivers supplying large parts of India, Bangladesh and China with water.

The effects mentioned concern primary biogeophysical systems that regulate the conditions on Earth that are vital for human life. They are not 'goods' in any ordinary sense of the word. If anything, they are factors of production or common-pool resources – sometimes producing what we term as 'ecosystem services' to humanity. We need therefore to use the word 'price' carefully. In many cases, there are no prices (indeed the lack of pricing for carbon dioxide disposal in the atmosphere can be seen as a root cause of the problem). On the other hand, it is also clear that we are facing an increased scarcity, and in this sense, a number of shadow 'values' are or should be increasing.

THE RAMSEY RULE GENERALIZED TO TWO SECTORS

To be more formal, the derivation of discounting in simple economic textbooks usually builds on a model of intertemporal optimization of the type

$$W = \int_0^T e^{-\rho t} U[C(t)]dt,$$

where $C(t)$ is consumption at time t and U is a measure of utility. The trade-offs between consumption at different points of time are given by two factors: the 'utility discount rate' ρ, and by the concavity of the utility function U, which measures 'inequality aversion': in a situation with economic growth, the future is thus given lower weight, the more concave U is.

With a concave utility function, U' is declining over time when consumption is growing, so that both terms in this expression are positive for this case. It is often assumed that the utility function has the simple form:

$$U(C) = \frac{1}{1-\alpha}C^{1-\alpha} \text{ for } \alpha > 0 \text{ and } U(C) = \text{Ln}C \text{ for } \alpha = 1. \quad (4.2)$$

This specification has the advantage that the elasticity of utility with respect to consumption is constant. In this case, the appropriate discount rate r is often called the 'Ramsey' rate:

$$r(t) = \rho + \alpha g_C(t), \quad (4.3)$$

where $g_C(t)$ is the growth rate of consumption. This discount rate will be constant over time if and only if the growth rate is constant. If we believe that growth rates will fall due to 'limits to growth', then discount rates would also fall over time (see, for example, Azar and Sterner, 1996).

In the debate on limits to growth and sustainability, the 'pessimists' assert that eternal growth is 'obviously' impossible due to limited resources on a finite planet. 'Optimists' (to which economists nowadays regularly belong), however, point to technology and new sectors as sources of growth. Communication and computing are clearly two examples of phenomenal economic growth that use few scarce natural resources.

I would thus tend to reject the notion of limited economic growth. But, on the other hand, the arguments about a finite globe obviously carry some weight. If future growth is concentrated only in some sectors (that use little or no material resources) while others have constant (or even

diminishing) levels, then this growth implies a changing output composition and presumably rising prices in the sectors that do not grow. This
brings us to the heart of the subject matter of this chapter.

If discounting, and hence growth, is essential for the valuation of future
damage to environmental systems, and if this growth is highly differential
between sectors, then this should be modelled explicitly. Let us build a
two-sector model with E to represent some aggregate measure of the environmental quality in society, while C is an aggregate measure of all other
goods. The utility function (4.2) will be replaced by $U = U(C, E)$, and we
will keep the constant elasticity of utility formulation from (4.2), but insert
a constant elasticity of substitution (CES) kernel to account for the interaction between C and E. This gives us the utility function where σ is the
elasticity of substitution:

$$U(C, E) = \frac{1}{1 - \alpha}[(1 - \gamma)C^{1 - (1/\sigma)} + \gamma E^{1 - (1/\sigma)}]^{\frac{(1 - \alpha)\sigma}{\sigma - 1}}. \qquad (4.4)$$

Going through the exact analogue to the traditional procedure outlined
above, the discount rate r is changed from (4.3) to (4.5) (see Hoel and
Sterner, 2007, or Guesnerie, 2004; for a more general analysis, see Gollier,
2007):

$$r = \rho + \left[(1 - \gamma^*)\alpha + \gamma^*\frac{1}{\sigma}\right]g_C + \left[\gamma^*\left(\alpha - \frac{1}{\sigma}\right)\right]g_E. \qquad (4.5)$$

It will be readily appreciated that (4.5) is a generalization of (4.3) with
the two sectors growing at growth rates g_C and g_E. The discount rate still
depends (as before) on the pure rate of time preferences ρ, but now on a
weighted average of the two growth rates. The weighting depends on the
elasticity of substitution and on γ^*, which can be interpreted as the 'value
share of environmental quality'.[4]

Note that the two discount rates coincide if either $\gamma^* = 0$ or $g_C = g_E$
(which basically means that there is no separate environmental sector)
or in a number of other circumstances, for instance, if the elasticities of
utility and substitution are unitary. Otherwise, the discount rate can be
either higher or lower. For the case of $g_C > g_E$, (4.5) gives a lower interest
rate than (4.3) if $\alpha\sigma > 1$. If both substitutability and utility curvatures are
small so that $\alpha\sigma < 1$, then interest rates will be higher. Note also that if
$\sigma \neq 1$ and $\alpha\sigma \neq 1$, the discount rate will not be constant over time even
if the growth rates for C and E are constant (since γ^* changes over time
when $\sigma \neq 1$).

As argued earlier, to calculate the future value of a change in environmental quality, we must consider both discounting and the change in the
relative price (or valuation) of the environmental quality. The valuation

of the environmental good is given by U_E/U_C. This fraction tells us by how much current consumption must increase to just offset a deterioration in current environmental quality of one unit (that is, to make current utility or well-being the same before and after the change in environmental quality and consumption). The relative change in this price is given by:

$$p = \frac{\dfrac{d}{dt}\left(\dfrac{U_E}{U_C}\right)}{\left(\dfrac{U_E}{U_C}\right)} = \frac{1}{\sigma}(g_C - g_E). \tag{4.6}$$

This price change will depend on the development over time of the two sectors in the economy C and E. The price change is positive provided that consumption increases relative to environmental quality over time and is larger, the smaller the elasticity of substitution. If, for example, the environmental quality is constant and consumption increases by 2.5 per cent a year and the elasticity of substitution is 0.5, this price will increase by a yearly rate of as much as 5 per cent.

The relative price effect will normally counteract the discounting effect. The combined effect of discounting and the relative price increase of environmental goods is given by $r - p$. If both r and p are positive, the sign of the combined effect is ambiguous. The reader may note that what I am here calling a 'combined effect' between relative prices and discounting could in another model be described in terms of a set of sector-specific discount rates. Mathematically they come to the same result, but I think the economic intuition benefits from thinking in terms of one discount rate ('the' rate at which we discount future costs and benefits) combined with changes in relative prices for goods in the future.

THE EFFECT OF RELATIVE PRICES FOR AN INTEGRATED ASSESSMENT OF CLIMATE DAMAGE

The Stern Review led to a heated debate with a large number of critical articles (see, for example, Dasgupta, 2006; Yohe, 2006; Nordhaus, 2007; Weitzman, 2007). William Nordhaus is generally considered one of the most prominent climate modellers, and his DICE model has to some extent acquired the status of being the 'standard' integrated assessment model (IAM). The Stern Review has been accused by Nordhaus, among others, of acquiring its radical results merely through the artifact of using low discount rates. Nordhaus uses the DICE model to reproduce results that are similar to those of the Stern Review – and quite different from those Nordhaus normally presents – simply by changing the discount rate used in the DICE model.

The issue of which discount rate to use is particularly complex and depends significantly on the context in which it is to be used. I shall not enter into the whole debate here but keep to the subject matter of this chapter by saying that if a relatively high rate is chosen – such as those preferred by Nordhaus – then this builds on the notion of continued high economic growth. Remembering that we are dealing in this case with a time horizon of several centuries, my point is that we must take into account that such growth would only be possible in some sectors. Other sectors will not grow, and some are doomed to decline – in particular because of climate change. This will imply significant changes in relative prices that ought to be taken into account.

Thomas Sterner and U. Martin Persson (2008) carry out such an analysis by slightly modifying Nordhaus's DICE model to contain two sectors. They amend the DICE model so that utility is dependent not only on material consumption goods, but also on environmental goods that are assumed to account initially for 10 per cent of utility. The environmental sector is assumed to have no inherent growth, but services from it are lowered through climate damage (the damage function is assumed to be quadratic in temperature). The authors show that this alternative approach – which builds on the same high discount rates as in Nordhaus – still yields results that are broadly in line with those of the Stern Review.

SUMMARY AND DISCUSSION

The use of cost–benefit analysis for such large, long-run and irreversible changes as climate change has been criticized on several grounds – among other things, that it is too partial an approach. One vital way in which it needs to be broadened is through allowing for the possibility that prices may change radically in the future. This chapter tries to take as its starting point a realistic view of the future: that cannot be one of constant, unwavering growth – equal for all sectors. Both logic and historic evidence show us that growth typically is concentrated in some sectors, determined by resource availability, technical innovations and the evolution of consumer preferences. Since discounting is so intimately tied to growth, we find that if sectoral growth is differential, then discounting needs to be complemented with a relative price change. This insight is not new (see, for example, Krutilla, 1967), but Michael Hoel and Thomas Sterner (2007) formalize it and show that relative price change and discounting can be deduced from the same coherent two-sector model framework. Sterner and Persson (2008) use this insight to discuss climate costs. Where Nordhaus (2007) criticizes the Stern Review for getting its high costs merely through the artifact of an arbitrarily low discount rate, Sterner and Persson show that very similar results can be obtained even with Nordhaus's higher discount rates, if only due consideration is taken of the price appreciation of scarce ecosystem services that will be damaged by future climate change.

The future value of ecosystem services is, of course, very hard to know. It depends on the interplay of at least three different sets of factors:

- technical parameters governing the development and interplay of ecosystem services;
- relative scarcity among sectors; and
- the evolution of preferences over time.

The first of these hinges on scientific knowledge that is as yet far from complete. This uncertainty is well analysed, for example, by Martin Weitzman in this book (see Chapter 5). This analysis leads to a separate and strong argument for precaution when it comes to potentially catastrophic effects of climate change. The second factor has been a main theme of this chapter: the fact that sectors will grow at very different growth rates will lead – due to the very long time frame involved – to large changes in sectoral composition. Depending on the crucial nature of the elasticities of substitution between the products and services from these different sectors, there may be considerable price effects. Finally, there is a third factor in the list above that will also have an important – but separate – effect on these relative prices: that is, our preferences (which may possibly evolve over time).

With rising income (and other societal changes over time), preferences for nature may change. It is often asserted that only the rich care for environmental 'amenities' and that the income elasticity of environmental services is thus larger than one. Against this backdrop, however, one may very well argue that it is the very poorest who are most dependent on environmental resources. If, for instance, the water goes bad on Manhattan Island, then (most of) the citizens will simply switch to drinking something else. It is painfully clear, in comparison, just how big the welfare losses are of poor water in, for example, Calcutta. This can be seen as an argument for a negative income elasticity of environmental services. My opinion is that both phenomena coexist. It is true, on the one hand, that people who have already secured the basic needs in life of food and protection can turn a greater portion of their attention to environmental amenities such as scenic beauty, altruistic and existence values of rare species in distant countries, and so on. On the other hand, it is also true that the poorest, who have little private capital, are the ones who are most heavily dependent on 'public' capital in the shape of common-pool resources that provide ecosystem services, such as water, firewood, fodder, building materials, medicinal herbs, small game and so on. They are simply different types of ecosystem service, and the issue of whether poor or rich have the higher share of utility from 'the' environment is partly a futile discussion that is better resolved by disaggregating 'the' environment into separate parts.

It is true that income elasticity effects such as those discussed in the preceding paragraph are important, in general, and for relative prices of environmental services, in particular. The effect discussed in this chapter is, however, conceptually a *separate* effect: the changes in relative price that, in my opinion, should be incorporated and that will dampen the effects of discounting are *not* the effect of (changing) tastes over time or income – but the result of scarcity from the supply side. Irrespective of our preferences, the value of productive coral reefs, land and water will rise if these become scarcer in the future – as is likely to be the effect of climate change.

NOTES

1. I would like to thank Martin Weitzman for several very astute comments on an earlier draft. Thanks also to Nick Stern, Jean-Philippe Touffut and the session participants of the Cournot Centre conference, 'The Economic Cost of Climate Change', held 18–19 December 2008 in Paris, for interesting discussions on the subject. Lastly, I thank the MISTRA programme CLIPORE for funding.
2. This chapter is based on the talk I gave at the Cournot Centre conference (2008), in which I was asked to speak after Michel Armatte (see Chapter 3). Michel provided an excellent overview of the development of modelling in the area of climate change, and he belongs specifically to these critics. This chapter was partly developed in response to him.
3. Furthermore, some private goods have emerged that are 'substitutes' for domestic labour, such as 'fast food'. Moreover, some 'domestic services' have been professionalized and are provided by public agencies or private firms (for example, pre-school care and dry cleaners).
4. γ^* is the share of total consumption expenditures that consumers would use on environmental quality if environmental quality was a good that could be purchased in the same manner as other consumption goods:

$$\gamma^* = \frac{(U_E/U_C)E}{[(U_E/U_C)E] + C} = \frac{\gamma E^{1-(1/\sigma)}}{(1-\gamma)C^{1-(1/\sigma)} + \gamma E^{1-(1/\sigma)}}.$$

REFERENCES

Azar, C. and T. Sterner (1996), 'Discounting and distributional considerations in the context of global warming', *Ecological Economics*, **19**, pp. 169–84.

Dasgupta, P. (2006), 'Comments on the Stern Review on the Economics of Climate Change', mimeo, University of Cambridge.

Gollier, C. (2007), 'Comment intégrer le risque dans le calcul économique', *Revue d'économie politique*, **117** (2), pp. 209–23.

Guesnerie, R. (2004), 'Calcul économique et développement durable', *La Revue Economique*, **55** (3), pp. 363–82.

Hoel, M. and T. Sterner (2007), 'Discounting and relative prices', *Climatic Change*, **84**, pp. 265–80.

IPCC (2007), *Climate Change 2007: Impacts, Adaptation and Vulnerability. Contribution of Working Group II to the Fourth Assessment Report of the Intergovernmental Panel on Climate Change*, edited by M.L. Parry, O.F. Canziani, J.P. Palutikof, P.J. van der Linden and C.E. Hanson, Cambridge: Cambridge University Press.

Krutilla, J.V. (1967), 'Conservation reconsidered', *American Economic Review*, **57** (4), pp. 777–86.

Nordhaus, W.D. (2007), 'A review of *The Stern Review on the Economics of Climate Change*', *Journal of Economic Literature*, **45**, pp. 686–702.

Stern, N. (2006), *The Economics of Climate Change: The Stern Review*, Cambridge: Cambridge University Press.

Sterner, T. and U.M. Persson (2008), 'An even Sterner review: introducing relative prices into the discounting debate', *Review of Environmental Economics and Policy*, **2** (1), pp. 61–76.

Weitzman, M.L. (2007), 'A review of *The Stern Review on the Economics of Climate Change*', *Journal of Economic Literature*, **45**, pp. 703–24.

Yohe, G. (2006), 'Some thoughts on the damage estimates presented in the Stern Review – an editorial', *Integrated Assessment Journal*, **6** (3), pp. 65–72.

5. Some basic economics of extreme climate change

Martin L. Weitzman

INTRODUCTION

Four big questions often asked about climate change are: (i) *how much* global warming and climate change will occur; (ii) *how bad* will it get; (iii) *when* will all this occur; and (iv) *what* should be done about it? This chapter attempts to explain why science and economics cannot resolve these questions to anywhere near the degree of accuracy that we have come to expect from more traditional applications of cost–benefit analysis (CBA), because there is so much deep structural uncertainty associated with climate change. The 'unknown unknowns' of climate change make CBA significantly more fuzzy in this arena than in more traditional applications, such as constructing roads, strengthening bridges or setting building codes in earthquake-prone zones. The chapter tries to make sense of this anomalous situation and explores what might be done in terms of actionable alternatives under such fuzzy circumstances.

Climate change is so complicated, and it involves so many sides of so many different disciplines and viewpoints, that no analytically tractable model or paper can aspire to illuminate more than a few facets of the problem. Because the problem is so complex, economists typically resort to numerical computer simulations. An integrated assessment model (IAM) for climate change is a multi-equation computerized model linking aggregate economic growth with simple climate dynamics in order to analyse the economic impacts of global warming. An IAM is essentially a dynamic model of an economy with a controllable greenhouse-gas-driven externality of endogenous greenhouse warming. IAMs have proven themselves useful for understanding several aspects of the economics of climate change – especially in describing outcomes from a complicated interplay of the very long lags and huge inertias involved.

A key starting point for any CBA of climate change should recognize that future temperatures or damages cannot be known exactly and must be expressed as a probability density function (PDF). Yet, most existing

IAMs treat central forecasts of temperatures or damages as if they were certain and then do some sensitivity analysis on parameter values. In the rare cases where an IAM formally incorporates uncertainty, it typically uses thin-tailed PDFs including, especially, truncation of PDFs at arbitrary cut-offs. (Often this truncation is more implicit than explicit, because a finite, discrete-point PDF is used.) What typically emerges from conventional IAM analysis is the so-called 'policy ramp' of gradually tightening emissions over time. The underlying rationale of the policy ramp is to postpone pain on climate-change prevention, because it is an investment whose pay-off comes only in the distant future (by human, if not geological, standards). When the distant-future pay-off times are considered, the rate of return on greenhouse gas (GHG) mitigation is lower than the rate of return on education, health, infrastructure, or a variety of other quicker-yielding public investments. As will be explained later, policy-ramp gradualism seems to be quite sensitive to the functional form of the assumed disutility of high-temperature changes, to how the extreme tail probabilities are specified, and to the rate of pure time preference used to discount future utilities and disutilities.

Modelling uncertain catastrophes presents some very strong challenges to economic analysis, the full implications of which have not yet been adequately confronted. CBA based on expected utility (EU) theory has been applied in practice primarily to cope with uncertainty in the form of a known thin-tailed PDF. I shall argue that the PDF of distant-future temperature changes is fat tailed. A thin-tailed PDF assigns a *relatively* much lower probability to rare events in the extreme tails than does a fat-tailed PDF.[1] (Even though both limiting probabilities are infinitesimal, the ratio of a thick-tailed probability divided by a thin-tailed probability approaches infinity in the limit.) Not much thought has gone into conceptualizing or modelling what happens to EU-based CBA for fat-tailed disasters. A CBA of a situation with known thin tails, even including whatever elements of subjective arbitrariness it might otherwise contain, can at least in principle make comforting statements of the generic form: 'if the PDF tails are cut off here, then EU theory will still capture and convey an accurate approximation of what is important'. Such accuracy-of-approximation, PDF-tail-cut-off statements, alas, do not exist in this generic sense for what in this chapter I am calling 'fat-tailed CBA'.

Fat-tailed CBA has strong implications that have been neither recognized in the literature nor incorporated into formal CBA modelling of disasters such as climate-change catastrophes. These implications raise many disturbing yet important questions, which will be dealt with somewhat speculatively in the concluding sections of this chapter. Partially answered questions and speculative thoughts aside, I contend that, at least in principle, fat-tailed CBA can change conventional thin-tail-based climate-change policy advice. This chapter argues that it is possible, and even numerically plausible, that the answers to the big policy question of what to do about climate change can hinge on the issue of how the high-temperature damages and tail probabilities are conceptualized and modelled. It is true that some reasonable-looking specifications and plausible parameter values can give rise to a gradualist policy ramp. I think it is also true, however, that some equally (or even more) reasonable specifications and parameter values can give very different results from a gradualist policy ramp. By implication, the advice coming out of conventional thin-tailed CBAs of climate change should be treated with caution until this low-probability, high-impact aspect is addressed seriously and resolved empirically in a true fat-tailed CBA.

This chapter explains in non-technical language a connection among the following four basic ideas: (i) the probability distribution of future global temperature has a fat tail at its upper extreme; (ii) the disutility damage of high temperatures is sensitive to the functional form that is assumed; (iii) when discounted at an uncertain rate of pure time preference that might be close to zero, the fat tails and temperature-sensitive disutilities can make expected present discounted damages very large; and (iv) because elevated stocks of CO_2 have such a long residence time in the atmosphere, and because it takes so long to learn about irreversible climate changes and to make mid-course corrections, a significant increase in expected welfare might be obtained if the upper extremes of the fat tail could be truncated before reaching catastrophic temperatures.

The final sections of the chapter concern the welfare and policy implications of coupling fat-tailed uncertainty with high disutility of extreme temperatures. Under any foreseeable technology, elevated stocks of CO_2 are committed to persist for a very long time in the atmospheric pipeline. It also takes a long time to learn about looming realizations of uncertain, but largely irreversible climate changes. Thus, CO_2 stock inertia, along with slow learning, makes it difficult to react to unfolding disasters by throttling back CO_2 flow emissions in time to avert an impending catastrophe. In this kind of situation, which is akin to trying to turn around an ocean liner in time to prevent a disaster, a large increase in expected welfare might be gained if some relatively benign form of fast geoengineering were deployable as an emergency, last-minute response for knocking down rapidly the bad fat tail of temperature change. Even if fast geoengineering is not a replacement for curtailing GHG emissions because it is too risky to be used as a mainline defence and it has too many other bad consequences, the logic of this chapter argues that it still might play an important niche role as an emergency-preparedness fallback component in a balanced portfolio of mixed options for dealing with climate change. The chapter highlights the idea that this aspect (the high expected value in this context of being able to truncate the bad fat tail quickly) may constitute a respectable economic underpinning supporting a well-funded research programme, undertaken now, to determine the feasibility, environmental side-effects, and cost effectiveness of fast geoengineering preparedness. The chapter concludes that, no matter what else is done realistically (within the realm of reason) to slow CO_2 build-ups, economic analysis lends some support to undertaking serious research now into the prospects of 'fast geoengineering preparedness' – as a state-contingent emergency option offering at least the possibility of knocking down catastrophic temperatures rapidly.

DEEP STRUCTURAL UNCERTAINTY ABOUT EXTREMES

In this section I try to make the case that standard CBAs or IAMs of climate change likely sidestep some important issues concerning improbable but extreme outcomes. I try to make this case by citing three aspects of the climate science that do not seem to be adequately covered by conventional economic analyses. While different aspects of structural uncertainty might additionally be cited, I restrict my case to these three examples, which I shall call 'Exhibits A, B and C'.

'Exhibit A' concerns the atmospheric level of GHGs over the last 800,000 years. Ice-core drilling in Antarctica began in the late 1970s and is still ongoing. The record of carbon dioxide (CO_2) and methane (CH_4) trapped in tiny ice-core bubbles was extended in 2008 to 800,000 years (see Lüthi et al., 2008). It is important to recognize that the numbers in this unparalleled 800,000-year record of GHG levels are among the very best data that exist in the science of paleoclimate. Almost all other data (including past temperatures) are inferred indirectly by proxy variables, whereas these ice-core GHG data are directly observed.

The pre-industrial-revolution level of atmospheric CO_2 (about two centuries ago) was 280 parts per million (ppm). The ice-core data show that carbon dioxide was never outside a range between 180 and 300 ppm during the last 800,000 years, with instances above 280 ppm exceedingly rare (to the point of being almost negligible). Currently, CO_2 is at 385 ppm. Methane was never higher than 750 parts per billion (ppb) in 800,000 years, but now this extremely potent GHG, which is 26 times more powerful than CO_2, is at 1780 ppb. Carbon-dioxide-equivalent (CO_2-e) GHGs are currently at 435 ppm. Even more alarming is the rate of change of GHGs, with increases in carbon dioxide never exceeding 30 ppm over any past thousand-year period, while now CO_2 has risen by 30 ppm in just the last 17 years.

Thus, anthropomorphic activity has elevated CO_2 and CH_4 to levels very far outside their natural range – and at a stupendously rapid rate. There is no analogue for anything like this having happened in the past geological record. Therefore, we do not really know with much confidence what will happen next. The link between GHG levels and temperature change in the ice-core record is not unicausal, and it is not fully understood, but this unsure link just adds more uncertainty to the picture. Any way one looks at it, GHGs are strongly implicated in global warming. Just to stabilize atmospheric CO_2 levels at twice pre-industrial-revolution levels would require not only stable but sharply *declining* emissions within a few decades from now. Forecasting ahead a century or two, the levels of atmospheric GHGs that may ultimately be attained (unless drastic measures are undertaken) have likely not existed for at least tens of millions of years, and the rate of change will likely be unique on a time scale of hundreds of millions of years.

Astonishingly, conventional CBAs and IAMs take almost no direct account of the magnitude of these unprecedented changes in GHGs – and the enormous uncertainty they create for an economic analysis of climate change. Perhaps even more remarkable is the fact that the gradualist policy ramp that emerges from standard CBAs and IAMs attains optimal stabilization at levels of CO_2 that are about 650–700 ppm within a century or two. This is my Exhibit A in the case that conventional CBAs and IAMs underplay, or sometimes even disregard, the tremendous structural uncertainties associated with climate change.

'Exhibit B' concerns the ultimate temperature response to such kind of unprecedented increases in GHGs.

So-called 'climate sensitivity' is a key macro-indicator of the *eventual* temperature response to GHG changes. Let $\Delta \ln CO_2$ be sustained relative change in atmospheric carbon dioxide, while ΔT is the equilibrium temperature response. Narrowly defined, climate sensitivity (here denoted S_1) converts $\Delta \ln CO_2$ into ΔT by the formula $\Delta T \approx (S_1 / \ln 2) \times \Delta \ln CO_2$. As the Intergovernmental Panel on Climate Change in its IPCC-AR4 (2007) Executive Summary puts it:

> The equilibrium climate sensitivity is a measure of the climate system response to sustained radiative forcing. It is not a projection but is defined as the global average surface warming following a doubling of carbon dioxide concentrations. It is *likely* to be in the range 2 to 4.5°C with a best estimate of 3°C, and is *very unlikely* to be less than 1.5°C. Values substantially higher than 4.5°C cannot be excluded, but agreement of models with observations is not as good for those values.

Climate sensitivity is not the same as temperature change, but for the benchmark-serving purposes of the simplistic example I shall be creating, I assume that the shapes of both PDFs are roughly similar after $\approx$ 100–200 years. The assumption is based on the fact that a doubling of anthropogenically injected CO_2-e GHGs relative to pre-industrial-revolution levels is unavoidable within the next $\approx$ 40–50 years, and those gases will plausibly remain well above 2 × pre-industrial levels for at least $\approx$ 100 + years thereafter. Other things being equal, higher values of climate sensitivity will produce higher temperatures at a more remote time in the distant future, which begs the question of whether enough can be learned sufficiently rapidly – relative to the super-long residence time of atmospheric CO_2 – to make meaningful mid-course corrections (and whether there would be sufficient political will to do it in time). To fully address these timing issues requires a more complete dynamic model (along with assumptions about the dynamics of information and learning), but I believe the example here is still telling.

In this chapter I am mostly concerned with the roughly 15 per cent of those S_1 'values substantially higher than 4.5°C' that 'cannot be excluded'. A grand total of 22 peer-reviewed studies of climate sensitivity published recently in reputable scientific journals and encompassing a wide variety of methodologies (along with 22 imputed PDFs of S_1) lie indirectly behind the above-quoted IPCC-AR4 (2007) summary statement. These 22 recent scientific studies are compiled in Table 9.3 and Box 10.2 of the IPCC-AR4. It might be argued that these 22 studies are of uneven reliability and their complicatedly related PDFs cannot easily be combined, but for the simplistic purposes of this illustrative example, I do not perform any kind of formal Bayesian model-averaging or meta-analysis (or even engage in informal cherry picking). Without question, a more sophisticated analysis of how to aggregate scientific data from different sources is required (including a more careful treatment of many aspects I am neglecting, such as possible multiplicative combining of probabilities from overlapping studies). Instead, I naively assume that all 22 studies have equal credibility, and for my purposes here their PDFs can be simplistically aggregated. The upper 5 per cent probability level averaged over all 22 climate-sensitivity studies cited in IPCC-AR4 is 7°C while the median is 6.4°C,[2] which I take as signifying approximately that $P[S_1 > 7°C] \approx 5\%$. Glancing at Table 9.3 and Box 10.2 of IPCC-AR4, it is apparent that the upper tails of these 22 PDFs tend to be sufficiently long and fat that one is allowed from a simplistically aggregated PDF of these 22 studies the rough approximation $P[S_1 > 10°C] \approx 1\%$. The empirical reason why these upper tails are long and fat dovetails beautifully with the theory of this chapter: inductive knowledge is always useful, of course, but simultaneously it is limited in what it can tell us about extreme events outside the range of experience – in which case one is forced back into depending more than one might wish on the prior PDF, which of necessity is largely subjective and relatively diffuse. As a recent *Science* commentary put it: 'Once the world has warmed by 4°C, conditions will be so different from anything we can observe today (and still more different from the last ice age) that it is inherently hard to say where the warming will stop' (Allen and Frame, 2007). However one looks at it, the long fat tail of climate sensitivity is disturbing. This is Exhibit B in my case that conventional CBAs and IAMs may not adequately cover the deep structural uncertainties associated with climate change.

'Exhibit C' concerns possibly disastrous releases over the long run of bad-feedback components of the carbon cycle that are currently omitted from most general circulation models of climate change. The chief worry here is a significant supplementary component that conceptually should be added on to climate sensitivity S_1. This omitted component concerns the powerful self-amplification potential of greenhouse warming due to heat-induced releases of sequestered carbon. One example is the huge volume of GHGs currently sequestered in Arctic permafrost and other boggy soils (mostly as methane, a particularly potent GHG). A yet more remote possibility, which in principle should also be included, is heat-induced releases of the even vaster offshore deposits of CH_4 trapped in the form of hydrates (aka clathrates). Here, there is a decidedly non-zero probability of destabilized methane seeping into the atmosphere if water temperatures over the continental shelves warm just slightly. The amount of methane involved is huge, although it is not precisely known.[3] Most estimates place the carbon content of methane hydrate deposits at about the same order of magnitude as the sum total of all of traditional fossil fuel deposits that will ever be extracted and burned by humans. A CH_4 outgassing-amplifier process could potentially precipitate (over the very long run, to be sure) a cataclysmic high-positive-feedback warming. This real physical basis for a catastrophe scenario is my Exhibit C in the case that conventional CBAs and IAMs do not adequately cover the structural uncertainties associated with possible climate-change disasters. Other examples of a real physical basis for a disastrous outcome could be cited, but this one will do here.

The real physical possibility of endogenous heat-triggered releases at high temperatures of the enormous amounts of naturally sequestered GHGs is a good example of indirect carbon-cycle feedback-forcing effects that I think should be included in the abstract interpretation of a 'concept of climate sensitivity' that is relevant for this chapter. What matters for the economics of climate change is the reduced-form relationship between atmospheric stocks of *anthropogenically injected* CO_2-e GHGs and temperature change. Instead of S_1, which stands for 'climate sensitivity narrowly defined', I work throughout the rest of this chapter with S_2, which (abusing scientific terminology somewhat here) stands for a more abstract 'generalized climate-sensitivity-like multiplier parameter' that includes heat-induced feedbacks on the forcing from the above-mentioned releases of naturally sequestered GHGs, increased respiration of soil microbes, climate-stressed forests, and other weakenings of natural carbon sinks. The transfer from Δ ln [anthropogenically injected CO_2-e GHGs] to eventual ΔT is not linear (and is not even a true long-run equilibrium relationship), but for the purposes of this highly aggregated example, the linear approximation is good enough. This suggests that a doubling of anthropogenically injected CO_2-e GHGs causes (very approximately) ultimate temperature change $\Delta T \approx S_2$.

The main point here is that the PDF of S_2 has an even longer, even fatter tail than the PDF of S_1. A recent study by Margaret Torn and John Harte (2006) can be used to give some very rough idea of the relationship of the PDF of S_2 to the PDF of S_1. It is universally accepted that in the absence of any feedback gain, $S_1 = 1.2°C$. If g_1 is the conventional feedback gain parameter associated with S_1, then $S_1 = 1.2/(1 - g_1)$, whose inverse is $g_1 = (S_1 - 1.2)/S_1$. Torn and Harte estimated that heat-induced GHG releases add about 0.067 of gain to the conventional feedback factor, so that (expressed in my language) $S_2 = 1.2/(1 - g_2)$, where $g_2 = g_1 + 0.067$. (The 0.067 is only an estimate in a linearized formula, but it is unclear in which direction higher-order terms would pull the formula and even if this 0.067-coefficient were considerably lower my point would remain.) Doing the calculations, $P[S_1 > 7°C] = 5\% = P[g_1 > 0.828] = P[g_2 > 0.895]$ implies $P[S_2 > 11.5°C] = 5\%$. Likewise, $P[S_1 > 10°C] = 1\% = P[g_1 > 0.88] = P[g_2 > 0.947]$ implies $P[S_2 > 22.6°C] = 1\%$ and presumably corresponds to a scenario where CH_4 and CO_2 are outgassed on a large scale from degraded permafrost soils, wetlands and clathrates.[4] The effect of heat-induced GHG releases on the PDF of S_2 is *extremely* nonlinear at the upper end of the PDF of S_2 because, so to speak, 'fat tails conjoined with fat tails beget yet-fatter tails'.

Of course my calculations and the numbers above can be criticized, but (quibbles and terminology aside) I do not think most climate scientists would say that these calculations are fundamentally wrong in principle or that there exists a clearly superior method for generating rough estimates of extreme-impact tail probabilities. I assume for purposes of this simplistic example that $P[S_2 > 10°C] \approx 5\%$ and $P[S_2 > 20°C] \approx 1\%$, implying that anthropogenic doubling of CO_2-e eventually causes $P[\Delta T > 10°C] \approx 5\%$ and $P[\Delta T > 20°C] \approx 1\%$, which I take as my base-case tail estimates in what follows. These small probabilities of what amounts to huge climate impacts occurring at some indefinite time in the remote future are wildly uncertain, unbelievably crude ballpark estimates – most definitely *not* based on hard science. But the subject matter of this chapter concerns just such kind of situations, and my overly simplistic example in this case does not depend at all on precise numbers or specifications. To the contrary, the major point I am trying to make is that such numbers and specifications *must* be imprecise and that this is a significant part of the climate-change economic-analysis problem, whose strong implications have thus far largely been ignored.

Stabilizing anthropogenically injected CO_2-e GHG stocks at anything like twice pre-industrial-revolution levels looks now like an extremely ambitious goal, which would require sharply declining GHG emissions within a few decades. Given current trends in emissions, we shall attain such a doubling of anthropogenically injected CO_2-e GHG levels around the middle of the twenty-first century and will then go far beyond that amount unless drastic measures are taken starting soon. Projecting current trends in business-as-usual GHG emissions, a *tripling* of anthropogenically injected CO_2-e GHG concentrations would be attained relative to pre-industrial-revolution levels by early in the twenty-second century. Countering this effect is the idea that we just might begin one day to seriously cut back on GHG emissions (especially if we learn that a high-S_2 catastrophe is looming – although the extraordinarily long inertial lags in the commitment pipeline converting CO_2 emissions into temperature increases might severely limit this option). On the other hand, maybe currently underdeveloped countries such as China and India will develop and industrialize at a blistering pace in the future with even more GHG emissions and even fewer GHG emissions controls than have thus far been projected. Or, who knows, we might one day discover a revolutionary new carbon-free energy source or make a carbon-fixing technological breakthrough. Perhaps natural carbon-sink sequestration processes will turn out to be weaker (or stronger) than we thought. There is also the unknown role of climate engineering. The recent scientific studies behind my crude ballpark numbers could turn out to be too optimistic or too pessimistic – or I might simply be misapplying these numbers by inappropriately using values that are either too high or too low, and so forth and so on. For the purposes of this very crude example (aimed at conveying some very rough empirical sense of the fatness of global-warming tails), I cut through the overwhelming enormity of climate-change uncertainty and the lack of hard science about tail probabilities by sticking with the overly simplistic story that $P[S_2 > 10°C] \approx P[\Delta T > 10°C] \approx 5\%$ and $P[S_2 > 20°C] \approx P[\Delta T > 20°C] \approx 1\%$. I cannot know precisely what these tail probabilities are, of course, but no one can – and *that* is the point here. To paraphrase again the overarching theme of this example: the moral of the story does not depend on the exact numbers or specifications in this drastic oversimplification, and if anything, it is enhanced by the fantastic uncertainty of such estimates.

It is difficult to imagine what $\Delta T \approx 10$–$20°C$ might mean for life on Earth, but such high temperatures have not been seen for hundreds of millions of years, and such a rate of change over a few centuries would be unprecedented even on a time scale of billions of years. Global average warming of 10–20°C masks tremendous local and seasonal variation, which can be expected to produce temperature increases *much* greater than this at particular times in particular places. Because these hypothetical temperature changes would be geologically instantaneous, they would effectively destroy planet Earth as we know it. At a minimum, such temperatures would trigger mass species extinctions and biosphere ecosystem disintegration matching or exceeding the immense planetary die-offs associated in Earth's history with a handful of previous geoenvironmental mega-catastrophes. There exist some truly terrifying consequences of mean temperature increases ≈ 10–$20°C$, such as: disintegration of Greenland and at least the western part of the Antarctic ice sheets with dramatic raising of sea level by perhaps 30 metres or so; critically important changes in ocean heat transport systems associated with thermohaline circulations; complete disruption of weather, moisture and precipitation patterns at every planetary scale; highly consequential geographic changes in freshwater availability; regional desertification; and so forth and so on.

All of the above-mentioned horrifying examples of climate-change mega-disasters are incontrovertibly possible on a time scale of centuries. They were purposely selected to come across as being especially lurid in order to drive home a valid point. The tiny probabilities of nightmare impacts of climate change are all such crude ballpark estimates (and they would occur so far in the future) that there is a tendency in the literature to dismiss altogether these highly uncertain forecasts on the 'scientific' grounds that they are much too speculative to be taken seriously. In a classical-frequentist mindset, the tiny probabilities of nightmare catastrophes are so close to zero that they are highly statistically insignificant at any standard confidence level, and one's first impulse can understandably be to just ignore them or wait for them to become more precise. My main theme contrasts sharply with the conventional wisdom of *not* taking seriously extreme temperature change probabilities *because* such probability estimates are not based on hard science and are statistically insignificant. The exact opposite logic holds, because there is a Bayesian sense in which, other things being equal, the more speculative and fuzzy are the tiny subjective tail probabilities of extreme events, the less ignorable and the more serious is the impact on present discounted expected utility for a risk-averse agent.

When fed into an economic analysis, the great open-ended uncertainty about eventual mean planetary temperature change cascades into yet much greater, yet much more open-ended uncertainty about eventual changes in welfare. There exists here a very long chain of tenuous inferences fraught with huge uncertainties in every link: it begins with unknown base-case GHG emissions; then compounded by huge uncertainties about how available policies and policy levers will transfer into actual GHG emissions; compounded by huge uncertainties about how GHG-flow emissions accumulate via the carbon cycle into GHG-stock concentrations; compounded by huge uncertainties about how and when GHG-stock concentrations translate into global mean temperature changes; compounded by huge uncertainties about how global mean temperature changes decompose into regional temperature and climate changes; compounded by huge uncertainties about how adaptations to, and mitigations of, climate-change damages are translated into utility changes – especially at a regional level; compounded by huge uncertainties about how future regional utility changes are aggregated – and then how they are discounted – to convert everything into expected present value global welfare changes. The result of this immense cascading of huge uncertainties is a reduced form of truly stupendous uncertainty about the aggregate expected present discounted utility impacts of catastrophic climate change, which mathematically is represented by a very spread-out, very fat-tailed PDF of what might be called 'welfare sensitivity'.

Even if a generalized climate sensitivity-like scaling parameter such as S_2 could be bounded above by some big number, the value of 'welfare sensitivity' is effectively bounded only by some *very* big number representing something like the value of statistical civilization as we know it or maybe even the value of statistical life on Earth as we know it. *This* is the essential point of this simplistic motivating example. Suppose it were granted, for the sake of argument, that an abstract climate sensitivity-like scaling parameter such as S_2 might somehow be constrained at the upper end by some fundamental law of physics that assigns a probability of exactly zero to temperature change being above some critical physical constant instead of continuously higher temperatures occurring with continuously lower probabilities trailing off asymptotically to zero. Even granted such an upper bound on S_2, the essential point here is that the enormous unsureness about (and enormous sensitivity of CBA to) an arbitrarily imposed 'damages function' for high-temperature changes makes the relevant reduced-form criterion of welfare sensitivity to a fat-tailed generalized scaling parameter seem almost unbelievably uncertain at high temperatures – to the point of being essentially unbounded for practical purposes. This is my Exhibit C.

FAT TAILS, HIGH-TEMPERATURE DISUTILITIES, AND DISCOUNTING

Because the integral of a PDF is one, the PDF of catastrophic temperatures must decline to an asymptote of zero probability. Thus, extreme outcomes can happen, but their likelihood diminishes to zero as a function of how extreme is the output. The fact that extreme outcomes cannot be eliminated altogether, but are theoretically possible with some positive probability, is not unique to climate change. What is worrisome is not the fact that extreme tails are *long per se* (reflecting the fact that a meaningful upper bound does not exist), but that they are *fat* (with probability density). The critical question is *how fast* does the probability of a catastrophe decline relative to the scope and impact of the catastrophe. Other things being equal, a thin-tailed PDF is of less concern, because the probability of the bad event declines exponentially (or faster). A fat-tailed distribution, where the probability declines polynomially in the temperature, can be much more worrisome.

A variety of mechanisms will produce a fat-tailed distribution of long-run temperature changes. One mechanism concerns the fact that climate sensitivity is of the form $S = 1/(1 - g)$ where $g < 1$ is a feedback-gain coefficient. If the PDF of g allows values near one, even with very low probability, then the PDF of S tends to be long and fat tailed.[5]

Another mechanism for generating fat tails is structural uncertainty.[6] The basic idea behind this mechanism can be illustrated by a specific example. Oversimplifying enormously, how warm the climate ultimately gets is approximately a product of two factors – anthropogenically injected CO_2-e GHGs and a critical climate sensitivity-like scaling multiplier. Both factors are uncertain, but the scaling parameter is more open-ended on the high side with a longer and fatter upper tail. This critical scale parameter reflecting huge scientific uncertainty is then used as a multiplier for converting aggregated GHG emissions – an input mostly reflecting economic uncertainty – into eventual temperature changes. Suppose the true value of this scaling parameter is unknown because of limited past experience, a situation that can be modelled *as if* inferences must be made inductively from a finite number of data observations. At a sufficiently high level of abstraction, each data point might be interpreted as representing an outcome from a particular scientific or economic study. Having an uncertain scale parameter in such a set-up can add a significant tail-fattening effect to posterior-predictive PDFs, even when Bayesian learning takes place with arbitrarily large (but finite) amounts of data. Loosely speaking, the driving mechanism here is that the operation of taking 'expectations of expectations' or 'probability distributions of probability distributions' spreads apart and fattens the tails of the reduced-form compounded posterior-predictive PDF. It is inherently difficult to learn from finite samples alone enough about the probabilities of extreme events to thin down the bad tail of the PDF, because, by definition, we do not get many data-point observations of such catastrophes. This mechanism provides some kind of a generic argument for why fat tails are almost inherent in many situations.

Although the basic idea is more general, it can be illustrated concretely by the relationship between the normal distribution and the Student-t. A normal distribution is thin tailed because the tail probabilities in the PDF decline faster than exponentially. If we do not know the parameters of the normal distribution (the mean and, more importantly, the standard deviation), but we have n observations drawn from the normal distribution, the implied posterior-predictive distribution is Student-t with n degrees of freedom. A Student-t PDF with n degrees of freedom is thick tailed because it is readily confirmed that the tails are polynomial of order n.

A Student-t 'child' posterior-predictive PDF from a large number of observations looks almost exactly like its bell-shaped normal 'parent' except that the probabilities are somewhat more stretched out, making the tails appear relatively fatter at the expense of a slightly flatter centre. In the limit, the ratio of the fat Student-t tail probability divided by the thin normal tail probability approaches infinity, even while both tail probabilities are approaching zero. Intuitively, a normal density 'becomes' a Student-t from a tail-fattening spreading apart of probabilities caused by the variance of the normal having itself a (inverted gamma) probability distribution. It is then no surprise that people are more averse qualitatively to a relatively fat-tailed, Student-t, posterior-predictive child distribution than they are to the relatively thin-tailed normal parent that begets it. Perhaps more surprising is the quantitative strength of this endogenously derived aversion to the effects of unknown tail structure. The story behind this quantitative strength is that fattened posterior-predictive bad tails represent structural or deep uncertainty about the possibility of rare high-impact disasters that – using colourful language here – 'scare' any agent having a utility function with relative risk aversion everywhere bounded above zero.

Uncertain structural parameters (coupled with finite data, under conditions of everywhere positive, relative risk aversion) can have strong consequences for CBA when catastrophes are theoretically possible, because in such circumstances it can drive CBA much more than anything else, including discounting. When fat-tailed temperature PDFs are combined with a utility function that is sensitive to high temperatures, it can make a difference on the outcomes of CBA. To see why the functional form of damages from high temperatures can be critical in this context, even when limited to a quadratic expression, consider the following formulation. Let $U(C, A)$ stand for utility as a function of consumption C and environmental amenities A. Let $A(\Delta T)$ stand for environmental amenities as a function of temperature change. Suppose in what follows, the base case

$$A(\Delta T) = \frac{1}{1 + \gamma (\Delta T)^2} \tag{5.1}$$

for some positive constant γ.

All existing IAMs treat high-temperature damages by a rather casual extrapolation of whatever specification is (typically arbitrarily) assumed to be the low-temperature 'damages function'. High-temperature damages extrapolated from a low-temperature damages function seem to be remarkably sensitive to assumed functional forms and parameter choices. Almost any function can be made to fit the low-temperature damages assumed by the modeller, even though these functions can give enormously different values at higher temperatures. Most IAM damages functions reduce welfare-equivalent consumption by a quadratic-polynomial multiplier equivalent to (5.1), with γ calibrated to some postulated loss for $\Delta T \approx$ 2–3°C. The standard conventional functional form combining consumption and environmental amenities is:

$$U(C, A) = \frac{(CA)^{1-\eta}}{1 - \eta}, \tag{5.2}$$

where η is the coefficient of relative risk aversion. The particular (and, in this context, perhaps peculiar) choice of functional form (5.1), (5.2) allows the economy to substitute consumption for high temperatures relatively easily, since the elasticity of substitution between C and A in this particular formulation is one.

There was never any more compelling rationale for the particular functional form (5.1), (5.2) than the comfort that economists feel from having worked with it before. In other words, the multiplicative quadratic-polynomial specification is extrapolated to assess climate-change damages at high temperatures for no better reason than casual familiarity with this particular form from other cost-of-adjustment dynamic economic models, where it has been used primarily for analytical simplicity in a situation that, at best, approximates reality for small changes. I would argue that if, for some unclear reason, climate-change economists want dependence of damages to be a multiplicative function of $(\Delta T)^2$ of form

(5.2), then a far better function at high temperatures for a consumption-reducing, welfare-equivalent, quadratic-based multiplier is the exponential form $A(\Delta T) = \exp[-\gamma(\Delta T)^2]$. Why? Look at the specification choice abstractly.

With isoelastic utility, the exponential specification is equivalent to $dU/U \propto dA$, while the polynomial specification is equivalent to $dU/U \propto dA/A$. For me it is obvious that, between the two, the former is much superior to the latter. When temperatures are already high in the latter case, why should the impact of dA on dU/U be artificially and unaccountably diluted by dividing dA by high values of A? The same argument applies to any polynomial in ΔT. I cannot prove that my favoured choice is the more reasonable of the two functional forms for high ΔT (although I truly believe that it is), but no one can disprove it either – and this is the point here.

The value of γ required for calibrating welfare-equivalent consumption at $\Delta T \approx$ 2–3°C to be (say) $\approx$ 97–98% of consumption at $\Delta T = 0$°C is so miniscule that both the polynomial-quadratic multiplier $1/[1 + \gamma(\Delta T)^2]$ and the exponential-quadratic multiplier $\exp[-\gamma(\Delta T)^2]$ give virtually identical outcomes for relatively small values of ΔT, but at ever higher temperatures they gradually, yet ever increasingly, diverge. With a fat-tailed PDF of ΔT, there can be a big difference between these two functional forms in the implied willingness to pay (WTP) to avoid or reduce uncertainty in ΔT. When the consumption-reducing, welfare-equivalent damages multiplier has the exponential form $\exp[-\gamma(\Delta T)^{\prime 2}]$, then with fat tails the WTP to avoid (or even reduce) fat-tailed uncertainty can be a very large fraction of consumption.

In a recent article, Thomas Sterner and U. Martin Persson (2008) tested on a leading IAM a utility function of the constant-elasticity-of-substitution (CES) form:

$$U(C, A) = \frac{1}{1-\eta}\left(C^{\frac{\sigma-1}{\sigma}} + bA^{\frac{\sigma-1}{\sigma}} \right)^{\frac{(1-\eta)\sigma}{\sigma-1}}, \tag{5.3}$$

where b and σ are positive constants, with σ being the (constant) elasticity of substitution between C and A. The particular multiplicative form (5.2) is a special case (for $\sigma = 1$) of the more general CES form (5.3). Sterner and Persson chose as elasticity of substitution what they argue is a more appropriate (than $\sigma = 1$) value of $\sigma = \frac{1}{2}$, and show that this change can make a big difference on the economic policy recommended by an IAM. When $\sigma = \frac{1}{2}$, the policy-ramp tightening of GHG emissions is much stronger and steeper than for the conventional IAM case $\sigma = 1$. When $\sigma = \frac{1}{2}$ and $\eta = 2$, which are the numerical values chosen by Sterner and Persson in their example, then (5.3) (with (5.1)) becomes:

$$U = -\left[\frac{1}{C} + \gamma(\Delta T)^2\right] - 1. \tag{5.4}$$

Exactly the same policy implications that are shown by Sterner and Persson would come out of the simpler specification of additively separable utility of the form:

$$U = \frac{C^{1-\eta}}{1-\eta} - \gamma(\Delta T)^2 \tag{5.5}$$

for the plausible coefficient of relative risk-aversion value $\eta = 2$. Form (5.5) is a standard isoelastic utility of consumption minus a conventional quadratic loss function in temperature changes. Because (5.4) and (5.5) are identical (except for a multiplicative constant), the numerical experiment of Sterner and Persson can be interpreted as showing that there is a big difference in policy implications between the standard (in the literature) multiplicative form (5.2) and the less standard (but no less familiar) additive form (5.5). In an optimal policy, the additive form (5.5) induces a much more stringent curtailment of GHG emissions than the multiplicative form (5.2). This demonstrates clearly how seemingly minor changes in the specification of high-temperature damages (here from multiplicative to additive) can dramatically change the climate-change policies recommended by an IAM. Such fragility of policy to forms of disutility functions might be considered a fourth exhibit – 'Exhibit D' if you will – in making the case that conventional CBAs and IAMs do not adequately cope with structural uncertainty – here uncertainty about the specification of damages.

There will be a *really* big impact on making optimal GHG-emissions policy more stringent if the discount rate used for discounting future climate-change disutilities is small (combined with fat tails of high temperatures and a disutility function that is sensitive to high-temperature damages). It is critical to bear in mind that the number being discussed here for discounting is the so-called 'rate of pure time preference' or 'utility discount rate', which is a subjective taste-like parameter that is difficult to pin down. It is *much* harder to argue that this utility discount rate should *not* be almost zero than it is to make such an argument for the so-called 'goods interest rate', which is far more directly tied to observed market rates of return on capital that are unquestionably significantly positive. (Any goods interest rate can be made compatible with a zero rate of pure time preference just by adjusting the elasticity of marginal utility appropriately.[7]) If a near-zero rate of pure time preference is used to discount disutilities of temperature damages, the optimal policy may curtail GHG emissions very severely in formulation (5.5) (or (5.4)). Even when the rate of pure time preference is positive, but it is not known, there can still be a big impact if there is a possibility of the rate of time preference being near zero. When this 'utility discount rate' itself has a PDF with non-negligible probability density in a neighbourhood of zero, then the expected present discounted disutility (from additive quadratic temperature damages) can be very large. In order to avoid such a probability-weighted bad possibility, the optimal policy will curtail GHG emissions, typically severely.

Reasonable attempts to constrict the length or the fatness of the 'bad' tail (or to modify the utility function) can still leave us with uncomfortably big numbers whose exact value depends non-robustly on artificial constraints or parameters that we really do not understand. The only legitimate way to avoid this potential problem is when there exists strong a priori knowledge that restrains the extent of total damages. If a particular type of idiosyncratic uncertainty affects only one small part of an individual's or a society's overall portfolio of assets, exposure is naturally limited to that specific component and bad-tail fatness is not such a paramount concern. Nevertheless, some very few but very important real-world situations have potentially *unlimited* exposure due to structural uncertainty about their potentially open-ended catastrophic reach. Climate change potentially affects the whole worldwide portfolio of utility by threatening to drive all of planetary welfare to disastrously low levels in the most extreme scenarios.

The part of the distribution of possible future outcomes that can most readily be learned (from inductive information of a form as if conveyed by data) concerns the relatively more likely outcomes in the middle of the distribution. From previous experience, past observations, plausible interpolations or extrapolations, and the law of large numbers, there may be at least some modicum of confidence in being able to construct a reasonable picture of the central regions of the posterior-predictive PDF. As we move towards probabilities in the periphery of the distribution, however, we are increasingly moving into the unknown territory of subjective uncertainty where our probability estimate of the probability distributions themselves becomes increasingly diffuse, because the frequencies of rare events in the tails cannot be pinned down by previous experiences or past observations. It is not possible to learn enough about the frequency of extreme tail events from finite samples alone to make the outcome of a CBA independent of artificially imposed bounds on the extent of possibly ruinous disasters. Climate-change economics generally – and the fatness of climate-sensitivity tails specifically – are prototype examples of this principle, because we are trying to extrapolate inductive knowledge far outside the range of limited past experience.

SOME IMPLICATIONS OF 'FAT-TAILED LOGIC'

By 'fat-tailed logic' I mean a combination of fat tails and temperature-sensitive disutilities, along with low rates of pure time preference. A common reaction to the conundrum for CBA implied by fat-tailed logic is to acknowledge its mathematical foundation, but to wonder how it is to be used constructively for deciding what to do in practice. Is this fat-tailed logic an economics version of an impossibility theorem, signifying that there are fat-tailed situations where economic analysis is up against a strong constraint on the ability of any quantitative analysis to inform us without committing to an empirical CBA framework that is based on some explicit numerical estimates of the miniscule probabilities of all levels of catastrophic impacts down to absolute disaster? Even if it were true that this logic represents a valid economic–statistical, precautionary-like principle that, at least theoretically, might dominate decision making, would not putting into practice this 'generalized precautionary principle' freeze all progress if taken too literally? Considering the enormous inertias that are involved in the build-up of GHGs, and the warming consequences, is the possibility of learning and mid-course corrections a plausible counter-weight to this fat-tailed logic, or, at the opposite extreme, has the commitment of GHG stocks in the ultra-long pipeline already fattened the bad tail so much that it makes little difference what is done in the near future about GHG emissions? How should the bad fat tail of climate uncertainty be compared with the bad fat tails of various proposed solutions, such as nuclear power, geoengineering, or carbon sequestration in the ocean floor? Other things being equal, this fat-tailed logic suggests as a policy response to climate change a relatively more cautious approach to GHG emissions, but *how much* more caution is warranted?

I simply do not know the full answers to the extraordinarily wide range of legitimate questions that fat-tailed logic raises. I do not think anyone does. But I also do not think that such questions can be allowed in good conscience to be simply shunted aside by arguing, in effect, that when probabilities are small and *im*precise, then they should be set *precisely* to zero. To the extent that uncertainty is formally considered at all in the economics of climate change, the artificial practice of using thin-tailed PDFs – especially the usual practice of imposing *de minimis* low-probability threshold cut-offs that casually dictate what part of the high-impact bad tail is to be truncated and discarded from CBA – seems arbitrary and problematic.[8] In the spirit that the unsettling questions raised by fat-tailed CBA for the economics of climate change must be addressed seriously, even while admitting that we do not know all of the answers, I offer here some speculative thoughts on what it all means. Even if the quantitative magnitude of what fat-tailed logic implies for climate-change policy seems somewhat hazy, the qualitative direction of the policy advice is nevertheless clear.

In ordinary limited-exposure or thin-tailed situations, there is at least the underlying theoretical reassurance that finite cut-off-based CBA might (at least in principle) be an arbitrarily close *approximation* to something that is accurate and objective. In fat-tailed, unlimited-exposure situations, by contrast, there is no such theoretical assurance underpinning the arbitrary cut-offs – and CBA outcomes are not robust to fragile assumptions about the likelihood of extreme impacts and how much disutility they cause.

One does not want to abandon lightly the ideal that CBA should bring independent empirical discipline to any application by being based on empirically reasonable parameter values. Even when fat-tailed logic applies, CBA based on empirically reasonable functional forms and parameter values might reveal useful information. Simultaneously one does not want to be obtuse by insisting that the catastrophe logic behind fat tails makes no practical difference for CBA, because the parameters just need to be determined empirically and then simply plugged into the analysis along with some extrapolative guesses about the form of the 'damages function' for high-temperature catastrophes (combined with speculative extreme-tail probabilities). So some sort of a tricky balance is required between being overawed by fat-tailed catastrophe logic into abandoning CBA altogether and being underawed into insisting that it is just another empirical issue to be sorted out by business-as-usual CBA. By all means plug in tail probabilities, disutilities of high impacts, rates of pure time preference, and so forth, and then see what emerges empirically – but do not be surprised when CBA outcomes are very sensitive to specifications and parameter values.

The degree to which the kind of 'generalized precautionary principle' that comes out of fat-tailed reasoning is relevant for a particular application must be decided on a case-by-case, 'rule of reason' basis. In the particular application to the economics of climate change, with so obviously limited data and limited experience about the catastrophic reach of climate extremes, to ignore or suppress the significance of rare fat-tailed disasters is to ignore or suppress what economic–statistical decision theory seems to be telling us here, which is potentially the most important part of the analysis.

Where does global warming stand in the portfolio of extreme risks currently facing us? There exist maybe half a dozen or so serious 'nightmare scenarios' of environmental disasters perhaps comparable in conceivable worst-case impact to catastrophic climate change. These might include: biotechnology, nanotechnology, asteroids, strangelets, pandemics, runaway computer systems and nuclear proliferation.[9] It may well be that each of these possibilities of environmental catastrophe deserves its own CBA application of fat-tailed logic along with its own empirical assessment of how much probability measure is in the extreme tails. Even if this were true, however, it would not lessen the need to reckon with the strong potential implications of fat-tailed logic for CBA in the particular case of climate change.

Perhaps it is little more than raw intuition, but for what it is worth I do not feel that the handful of other conceivable environmental catastrophes are nearly as critical as climate change. I illustrate with two specific examples. The first is widespread cultivation of crops based on genetically modified organisms (GMOs). At casual glance, climate-change catastrophes and bio-engineering disasters might look similar. In both cases, there is deep unease about artificial tinkering with the natural environment, which can generate frightening tales of a planet ruined by human hubris. Suppose for specificity that with GMOs the overarching fear of disaster is that widespread cultivation of so-called 'Frankenfood' might somehow allow bioengineered genes to escape into the wild and wreak havoc on delicate ecosystems and native populations (including, perhaps, humans), which have been fine-tuned by millions of years of natural selection. At the end of the day I think that the potential for environmental disaster with Frankenfood is much less than the potential for environmental disaster with climate change – along the lines of the following loose and oversimplified reasoning.

In the case of Frankenfoods interfering with wild organisms that have evolved by natural selection, there is at least *some* basic underlying principle that plausibly dampens catastrophic jumping of artificial DNA from cultivars to landraces. After all, nature herself has already tried endless combinations of mutated DNA and genes over countless millions of years, and what has evolved in the fierce battle for survival is only an infinitesimal subset of the very fittest permutations. In this regard there exists at least some inkling of a prior argument making it fundamentally implausible that Frankenfood artificially selected for traits that humans find desirable will compete with or genetically alter the wild types that nature has selected via Darwinian survival of the fittest. Wild types have already experienced innumerable small-step genetic mutations, which are perhaps comparable to large-step, human-induced artificial modifications, and which have not demonstrated survival value in the wild. Analogous arguments may also apply for invasive 'superweeds', which so far represent a minor cultivation problem lacking ability to displace either landraces or cultivars. Besides all this, safeguards in the form of so-called 'terminator genes' can be inserted into the DNA of GMOs, which directly prevent GMO genes from reproducing themselves.

A second possibly relevant example of comparing climate change with another potential catastrophe concerns the possibility of a large asteroid hitting Earth. In the asteroid case, it seems plausible to presume that there is much more inductive knowledge (from knowing something about asteroid orbits and past collision frequencies) pinning down the probabilities to very small 'almost known' values. If we use $P[\Delta T > 20°C] \approx 1\%$ as the very rough probability of a climate-change cataclysm occurring within the next two centuries, then this is roughly 10,000 times larger than the probability of a large asteroid impact (of a one-in-a-hundred-million-years size) occurring within the same time period.

Contrast the above discussion about plausible magnitudes or probabilities of disaster for genetic engineering or asteroid collisions with possibly catastrophic climate change. The climate-change 'experiment', whose eventual outcome we are trying to infer now, 'tests' the planet's response to a geologically instantaneous exogenous injection of GHGs. An exogenous injection of this much GHGs this fast seems unprecedented in Earth's history stretching back perhaps billions of years. Can anyone honestly say now, from very limited prior information and very limited empirical experience, what are reasonable upper bounds on the eventual global warming or climate change that we are currently trying to infer will be the outcome of such a first-ever planetary experiment? What we do know about climate science and extreme tail probabilities is that the rate of change of GHGs seems almost unprecedented in geological history; planet Earth hovers in an unstable trigger-prone 'whipsaw' ocean-atmosphere system;[10] chaotic dynamic responses to geologically instantaneous GHG shocks are possible; and all 22 recently published studies of climate sensitivity cited by IPCC-AR4 (2007), when mechanically aggregated together, estimate on average that $P[S_1 > 7°C] \approx 5\%$. To my mind this open-ended aspect with a way-too-high subjective probability of a catastrophe makes GHG-induced global climate change vastly more worrisome than cultivating Frankenfood or colliding with large asteroids.

These two examples hint at making a few meaningful distinctions among the handful of situations where fat-tailed logic might reasonably apply. My discussion here is hardly conclusive, so we cannot rule out a biotech or asteroid disaster. I would say, however, on the basis of this line of argument that such disasters seem *very very* unlikely, whereas a climate disaster seems 'only' *very* unlikely. In the language of this chapter, synthetic biology or large asteroids feel more like high-knowledge situations that we know a lot more about relative to climate change, which by comparison feels more like a low-knowledge situation about which we know relatively little. Whether my argument here is convincing or not, the overarching principle is this: the mere fact that my logic might also apply to a few other environmental catastrophes does *not* constitute a valid reason for excluding it from applying to climate change.

The simplistic story I am telling here represses the real-option value of waiting and learning. Concerning this aspect, however, with climate change we are on the four horns of two dilemmas. The horns of the first dilemma are the twin facts that built-up stocks of GHGs might end up *ex post* representing a hugely expensive, irreversible accumulation, but so too might massive investments in non-carbon technologies that are at least partly unnecessary.

The second dilemma is the following. Because climate-change catastrophes develop more slowly than some other potential catastrophes, there is ostensibly somewhat more chance for learning and mid-course corrections with global warming relative to, say, biotechnology (but not necessarily relative to asteroids when a good tracking system is in place). The possibility of 'learning by doing' may well be a more distinctive feature of global-warming disasters than some other disasters, and in that sense deserves to be part of an optimal climate-change policy. The other horn of this second dilemma, however, is the nasty fact that the ultimate climate response to GHGs has tremendous inertial pipeline-commitment lags of several centuries up to millennia (via the very long atmospheric residence time of CO_2). When all is said and done, I do not think that there is a smoking gun in the biotechnology, asteroid, or any other catastrophe scenario quite like the idea that a crude amalgamation of numbers from the most recent peer-reviewed published scientific articles is suggesting, something like $P[S_2 > 10°C] \approx 5\%$ and $P[S_2 > 20°C] \approx 1\%$.

The logic of catastrophic climate change seems to be suggesting here that the debate about what interest rate to use for discounting goods and services, which has dominated the discussion so far, may be secondary to a debate about the open-ended catastrophic reach of climate disasters. While it is always fair game to challenge the assumptions of a model, when theory provides a generic result (such as 'free trade is Pareto optimal' or 'steady growth eventually outstrips one-time change') the burden of proof is commonly taken as being upon whomever wants to over-rule the theorem in a particular application. The burden of proof in climate-change CBA might fall upon whomever calculates expected discounted utilities and disutilities without considering that structural uncertainty might matter more than discounting or pure objective risk.

POSSIBLE IMPLICATIONS FOR CLIMATE-CHANGE POLICY

Instead of the existing IAM emphasis on estimating or simulating economic impacts of the more plausible climate-change scenarios, to at least

compensate partially for finite-sample bias, the model of this chapter calls
for a dramatic oversampling of those stratified climate-change scenarios
associated with the most adverse imaginable economic impacts in the
bad fat tail. With limited sampling resources for the big IAMs, Monte
Carlo analysis could be used much more creatively – not necessarily to
defend a specific policy result, but to experiment seriously in order to find
out more about what happens with fat-tailed uncertainty and significant
high-temperature damages in the limit as the grid size and number of runs
increase simultaneously. Of course an emphasis on sampling climate-
change scenarios in proportion to utility-weighted probabilities of occurrence forces us to estimate subjective probabilities down to extraordinarily
tiny levels and also to put degree-of-devastation weights on disasters with
damage impacts up to perhaps being welfare-equivalent to losing 99 per
cent (or possibly even more) of consumption – but that is the price we must
be willing to pay for having a genuine economic analysis of potentially
catastrophic climate change.

In situations of potentially unlimited damage exposure like climate
change, it might be appropriate to emphasize a slightly better treatment of
the worst-case, fat-tail extremes – and what might be done about them, at
what cost – relative to refining the calibration of most-likely outcomes or
rehashing point estimates of discount rates (or climate sensitivity). A clear
implication of this chapter is that greater research effort is relatively inef-
fectual when targeted at estimating central tendencies of what we already
know relatively well about the economics of climate change in the more-
plausible scenarios. A much more fruitful goal of research might be to
aim at understanding even slightly better the deep uncertainties concerning the *less* plausible scenarios located in the bad fat tail. (Alas, the tails
are the very part of a PDF that is most difficult to learn, presenting yet
another policy dilemma.) I also believe that an important complementary
research agenda, which stems naturally from the analysis of this chapter,
is the crying need to comprehend much better *all* of the options for possibly dealing with high-impact climate-change extremes, without trying to
pre-censor any of them as socially unacceptable or politically incorrect.

When analysing the economics of climate change, perhaps it might
be possible to make 'back-of-the-envelope' comparisons with empirical
probabilities and mitigation costs for extreme events in the insurance
industry. One might try to compare numbers on, say, a homeowner buying
fire insurance (or buying fire-protection devices, or a young adult purchasing life insurance, or others purchasing flood-insurance plans) with
cost–benefit guesstimates of the world buying an insurance policy going
some way towards mitigating the extreme high-temperature possibilities.
On a US national level, rough comparisons could perhaps be made with

the potentially huge pay-offs, small probabilities, and significant costs involved in countering terrorism, building anti-ballistic missile shields, or neutralizing hostile dictatorships possibly harbouring weapons of mass destruction. A crude natural metric for calibrating cost estimates of climate-change environmental insurance policies might be that the United States already spends approximately $2\frac{1}{2}$ per cent of national income on the cost of a clean environment.[11] All of this having been said, the bind we find ourselves in now on climate change starts from a diffuse prior situation to begin with, and is characterized by extremely slow convergence of inductive knowledge towards resolving the tail uncertainties – relative to the lags and irreversibilities from not acting before structure is more fully identified.

The point of all of this is that economic analysis is not completely helpless in the presence of deep structural uncertainty and potentially unlimited exposure. We can say a few important things about the relevance of fat-tailed CBA to the economics of climate change. The analysis is much more frustrating and much more subjective – and it looks much less conclusive – because it requires some form of speculation (masquerading as an 'assessment') about the extreme bad-fat-tail probabilities and utilities. Compared with the thin-tailed case, CBA of fat-tailed potential catastrophes is inclined to favour paying a lot more attention to learning how fat the bad tail might be and – if the tail is discovered to be too heavy for comfort after the learning process – is a lot more open to at least considering undertaking serious mitigation measures (including, perhaps, geoengineering in the case of climate change) to slim it down fast. This paying attention to the feasibility of slimming down overweight tails is likely to be a perennial theme in the economic analysis of catastrophes. The key economic questions here are: what is the overall cost of such a tail-slimming weight-loss programme, and how much of the bad fat does it remove from the overweight tail?

AN ANALYTICAL FOUNDATION FOR FAST GEOENGINEERING?

The economist's case for a carbon tax is traditionally made without explicit reference to extreme tail behaviour. This argument is presumably strengthened when extreme tail events are considered. Whatever value it happens to be, the 'uncomfortably big number' for expected disutility that tends to emerge from fat-tailed logic can be reduced by imposing carbon taxes. So the very first thing to say here is that the fat upper tail of the PDF of possible temperature changes lends even greater urgency to reducing GHG emissions by levying a substantial tax on the burning of fossil fuels. Having said this, there is more to say. The fat tails introduce some distinctive issues of their own. Responsible economic analysis of fat tails implies some tolerance for at least considering extreme-sounding proposals that are not normally placed on the policy table for discussion. One consequence of fat-tailed logic might concern the role of fast-acting planetary geoengineering. The opinion that follows might be construed as editorializing, but it seems to me that the analysis of this chapter leads logically to a narrowly defined niche role for a reliable backstop technology that can effectively knock down high planetary temperatures quickly in case of emergency.

What I mean by 'fast geoengineering' is any action having the possibility to lower global temperatures quickly – within decades or even years. Practically, at this time fast geoengineering means albedo enhancement by injecting sunlight-reflective particulates or aerosols, such as sulphur dioxide precursors, into the stratosphere. I do not touch upon the science of fast geoengineering, and even tread lightly upon the economics.[12] My main focus here is on the narrow question of whether the analytical argument of this chapter supports a special niche role for fast geoengineering – as one important option in a balanced portfolio of global warming strategies and responses. I think the answer is a qualified yes.

The analysis of this chapter is suggesting that a significant component of the overall expected damages of climate change may be located in the fat upper tail of the temperature distribution. Cut out the fat upper tail, and you have cut out a major part of the expected disutility of global warming, goes the argument. According to this logic, a large increase in expected welfare might be gained if some relatively benign form of fast geoengineering were deployed in readiness to rapidly derail severe greenhouse heating – should this contingency materialize. Because of the largely irreversible long pipeline commitment of atmospheric CO_2, this argument might hold even though higher temperatures tend to materialize later, and the 'emergency' might unfold over a time scale of centuries.

Fast geoengineering seems quite risky, if for no other reason than the law of unintended consequences, and it cannot ward off all the bad effects of high atmospheric CO_2, such as ocean acidification. To say, however, that fast geoengineering does not now look like a panacea for all the effects of climate and atmosphere changes should not be to prejudge that it may not have a very important, perhaps even crucial, future role to play in a balanced portfolio of responsible climate-change policies. Even if fast geoengineering (Plan B) is not a replacement for curtailing GHG emissions (Plan A) – because it is too risky to be used as a mainline defence – it might still be critical to have a Plan-B option in reserve. The analysis of this chapter formalizes a possibly large potential welfare gain from having the capability to slim down quickly a bad, fat global-warming tail during a worst-case emergency. In my opinion, this appears to be a legitimate argument for a well-funded Plan-B research programme, undertaken now, which might include pilot studies and small-scale field testing. The purpose would be to determine the feasibility, environmental side-effects, and cost-effectiveness of responsible geoengineering preparedness – whose intended use is as a state-contingent option giving the ability to respond rapidly to a bad future realization of global-warming uncertainty.

A huge issue with fast geoengineering is that, as an externality, it has diametrically opposite cost properties from curtailing emissions of GHGs. For me, the two really inconvenient truths about climate change are: (i) CO_2 abatement is really costly; and (ii) fast geoengineering is really cheap. Like it or not, whether it is a panacea or not, whether it lulls the public into a false sense of security that undermines legitimate Plan-A GHG-curtailment strategies or not, the incredible economics of geoengineering simply cannot be ignored.[13] The fast geoengineering option currently looks so unbelievably inexpensive as a quick fix for extreme temperature changes that virtually any middle-power developed country might be tempted to implement it unilaterally. For me this means that – as well as there being a strong policy argument that *now* is the time to learn a lot more about fast geoengineering – there is an additional strong policy argument that *now* is also the time to start thinking seriously about an international framework governing the use of this scary option.

CONCLUSION

Heroic attempts at constructive suggestions notwithstanding, it is painfully apparent that fat-tailed logic makes economic analysis trickier and more open-ended in the presence of deep structural uncertainty. The economics of fat-tailed catastrophes raises difficult conceptual issues, which cause the analysis to appear less scientifically conclusive and to look more contentiously subjective than what comes out of an empirical CBA of more usual thin-tailed situations. But if this is the way things are with fat tails, then this is the way things are, and it is an inconvenient truth to be lived with rather than a fact to be evaded just because it looks less scientifically objective in cost–benefit applications.

Perhaps in the end, the climate-change economist can help most by not presenting a cost–benefit estimate as if it is accurate and objective for what is inherently a fat-tailed situation with potentially unlimited downside exposure. Perhaps they should not even present the analysis as if it is an approximation to something that is accurate and objective – but instead stress somewhat more openly the fact that such an estimate might conceivably be arbitrarily *in*accurate depending on what is subjectively assumed about the high-temperature damages function along with assumptions about the fatness of the tails and/or where they have been cut off. Simply acknowledging more openly the incredible magnitude of the deep structural uncertainties that are involved in climate-change analysis – and explaining better to policy makers that the artificial crispness conveyed by conventional IAM-based CBAs is especially and unusually misleading compared with more-ordinary non-climate-change CBA situations – might elevate the level of public discourse concerning what to do about global warming. All of this is naturally unsatisfying, frustrating, and not what economists are used to doing. In rare situations like climate change, however, where fat-tailed logic applies, we may be deluding ourselves and others with misplaced concreteness if we think that we are able to deliver anything much more precise than this with even the biggest and most-detailed climate-change IAMs as currently constructed and deployed.

This chapter has presented a basic theoretical principle that holds under temperature-sensitive disutilities and potentially unlimited exposure. In principle, what might be called the catastrophe-insurance aspect of such a fat-tailed, unlimited-exposure situation, which can never be fully learned away, can dominate discounting, objective-probability risk and consumption smoothing. Even if this principle in and of itself does not provide an easy answer to questions about how much catastrophe insurance to buy (or even an easy answer in practical terms to the question of what exactly *is* catastrophe insurance buying for climate change or other applications), I believe it still might provide a useful way of framing the economic analysis of catastrophes.

NOTES

1. As I use the term, a PDF has a 'fat' (or 'thick' or 'heavy') tail when its moment generating function (MGF) is infinite – that is, the tail probability approaches zero *more slowly* than exponentially. The standard example of a fat-tailed PDF is the power law (aka Pareto aka inverted polynomial) distribution, although, for example, a lognormal PDF is also fat tailed, as is an inverted normal or inverted gamma or Student-*t*. By this more or less standard definition, a PDF whose MGF is finite has a 'thin tail' – that is, the tail probability approaches zero *more rapidly* than exponentially. A normal or a gamma are examples of thin-tailed PDFs, as is *any* PDF having finite supports, such as a uniform distribution or a discrete-point distribution.
2. Details of this calculation are available upon request. Eleven of the studies in Table 9.3 overlap with the studies portrayed in Box 10.2. Four of these overlapping studies conflict on the numbers given for the upper 5 per cent level. For three of these differences, I chose the Table 9.3 values on the grounds that all of the Box 10.2 values had

been modified from the original studies to make them have zero probability mass above 10°C. (The fact that all PDFs in Box 10.2 have been normalized to zero probability above 10°C biases my upper 5 per cent averages here towards the low side.) With the fourth conflict, I substituted 8.2°C from Box 10.2 for the ∞ in Table 9.3 (which arises only because the method of the study itself does not impose any meaningful upper-bound constraint). The only other modification was to average the three reported volcanic-forcing values in Table 9.3 into one upper 5 per cent value of 6.4°C.

3. IPCC4-AR4 contains some discussion of methane releases.
4. I am grateful to John Harte for guiding me through these calculations, although he should not be blamed for how I am interpreting or using the numbers in what follows. The Torn and Harte study (2006) is based on an examination of the 420,000-year record from Antarctic ice cores of temperatures along with associated levels of CO_2 and CH_4. While based on different data and a different methodology, the study of Sheffer et al. (2006) supports essentially the same conclusions as Torn and Harte. A completely independent study from simulating an interactive coupled climate-carbon model of intermediate complexity in Matthews and Keith (2007) confirms the existence of a strong carbon-cycle feedback effect with especially powerful temperature amplifications at high climate sensitivities.
5. This idea is developed in an influential article by Roe and Baker (2007).
6. This idea is developed extensively in Weitzman (2009).
7. This is explained, for example, in Dasgupta (2007).
8. Adler (2007) sketches out in some detail the many ways in which *de minimis* low-probability-threshold cut-offs are arbitrary and problematic in more-ordinary regulatory settings.
9. Many of these are discussed in Posner (2004), Parson (2007) and Sunstein (2007).
10. On the nature of this unstable 'whipsaw' climate equilibrium, see Hansen et al. (2007).
11. US Environmental Protection Agency (1990), executive summary projections for 2000, which I updated and extrapolated to 2007.
12. Some of the science is reviewed in Rasch et al. (2008). The idea of fast geoengineering has been around for a long time, but was recently given much visibility by the influential article of Crutzen (2006).
13. Barrett (2008) contains an excellent discussion of some implications of what he has dubbed 'the incredible economics of geoengineering'.

REFERENCES

Adler, Matthew D. (2007), 'Why *de minimis*?', AEI-Brookings Joint Center Related Publication 07–17, Washington, DC.

Allen, Myles R. and David J. Frame (2007), 'Call off the quest', *Science*, **318**, pp. 582–83.

Barrett, Scott (2008), 'The incredible economics of geoengineering', *Environmental and Resource Economics*, **39**, pp. 45–54.

Crutzen, Paul J. Albedo (2006), 'Enhancement by stratospheric sulfur injections: a contribution to resolve a policy dilemma?', *Climatic Change*, **77**, pp. 211–20.

Dasgupta, Partha (2007), 'Commentary: the Stern Review's Economics of Climate Change', *National Institute Economic Review*, **199**, pp. 4–7.

Hansen, James, Makiko Sato, Pushker Kharechal, Gary Russell, David W. Lea and Mark Siddal (2007), 'Climate change and trace gases', *Philosophical Transactions of the Royal Society A*, **365** (1856), pp. 1925–54.

IPCC-AR4 (2007), *Climate Change 2007: The Physical Science Basis. Contribution of Working Group I to the Fourth Assessment Report of the Intergovernmental Panel on Climate Change*, edited by S. Solomon, D. Qin, M. Manning, Z. Chen,

M. Marquis, K.B. Averyt, M. Tignor and H.L. Miller, Cambridge, UK and New York, NY, USA: Cambridge University Press.

Lüthi, Dieter, Martine Le Floch, Bernhard Bereiter, Thomas Blunier, Jean-Marc Barnola, Urs Siegenthaler, Dominique Raynaud, Jean Jouzel, Hubertus Fischer, Kenji Kawamura and Thomas Stocker (2008), 'High-resolution carbon dioxide concentration record 650,000–800,000 years before present', *Nature*, **453**, pp. 379–82.

Matthews, H. Damon and David W. Keith (2007), 'Carbon-cycle feedbacks increase the likelihood of a warmer future', *Geophysical Research Letters*, **34**, L09702.

Parson, Edward A. (2007), 'The Big One: a review of Richard Posner's *Catastrophe: Risk and Response*', *Journal of Economic Literature*, **45** (March), pp. 147–64.

Posner, Richard A. (2004), *Catastrophe: Risk and Response*, Oxford: Oxford University Press.

Rasch, Philip J., Simone Tilmes, Richard P. Turco, Alan Robock, Luke Oman, Chih-Chieh Chen, Georgiy L. Stenchikov and Rolando R. Garcia (2008), 'An overview of geoengineering of climate using stratospheric sulfate aerosols', *Philosophical Transactions of the Royal Society A*, **366**, pp. 4007–37.

Roe, Gerard H. and Marcia B. Baker (2007), 'Why is climate sensitivity so unpredictable?', *Science*, **318**, pp. 629–32.

Sheffer, Martin, Victor Brovkin and Peter M. Cox (2006), 'Positive feedback between global warming and atmospheric CO_2 concentration inferred from past climate change', *Geophysical Research Letters*, **33** (10), L10702.

Sterner, Thomas and U. Martin Persson (2008), 'An even Sterner review: introducing relative prices into the discounting debate', *Review of Environmental Economics and Policy*, **2** (1), pp. 61–76.

Sunstein, Cass R. (2007), *Worst-Case Scenarios*, Cambridge, MA: Harvard University Press.

Torn, Margaret S. and John Harte (2006), 'Missing feedbacks, asymmetric uncertainties, and the underestimation of future warming', *Geophysical Research Letters*, **33** (10), L10703.

US Environmental Protection Agency (1990), *Environmental Investments: The Cost of a Clean Environment*, Washington, DC: US Government Printing Office.

Weitzman, Martin L. (2009), 'On modeling and interpreting the economics of catastrophic climate change', *Review of Economics and Statistics*, **91** (1), February, pp. 1–19.

6. Round table discussion: Economics and climate change – where do we stand and where do we go from here?

Inge Kaul, Thomas Schelling, Robert M. Solow, Nicholas Stern, Thomas Sterner and Martin L. Weitzman

Under the chairmanship of Robert Solow (MIT), this round table discussion brought together Inge Kaul (Hertie School of Governance, Berlin), Thomas Schelling (University of Maryland), Nicholas Stern (London School of Economics), Thomas Sterner (University of Gothenburg) and Martin Weitzman (Harvard University).

Robert Solow (MIT) This round table is intended as a wind-up discussion, not to summarize what has already been said, but perhaps to draw conclusions and express different points of view. The contributions so far have been lacking one common view of the matter: the conclusion, after a great deal of study and modelling and contemplation of alternatives, that it would be better not to take drastic policy action right away, even if that were possible; that the correct strategy may be to wait, acquire information, invest in physical and human capital, increase the capacity of the world economy to produce and to meet climate possibilities.[1] How does this difference of opinion arise? Should we regard the matter as settled? There has been more unanimity in the contributions of this volume than I think there is in the world at large, or in the part of the research world that is interested in this question. I don't think the differences of opinion arise out of differences in the fundamental science, but out of differences in the economic analysis. So I'm going to ask my colleagues to comment on that matter and then move to another question.

Nicholas Stern (London School of Economics) In models like DICE, the view of climate change is made up, in very crude terms, of two things: how

135

much you're bothered about the future (and in this case the future over a long time, because these effects come through very slowly) and, second, your view of the magnitude of the damages and risks. I think the DICE approach gets both things wrong.

The optimal path in the DICE model takes you above 700, 750 parts per million CO_2-equivalent some time around the end of this century moving into the next. What's involved there? If you managed to stabilize at 750 CO_2-equivalent (that is, CO_2 and the other greenhouse gases), you would have a roughly 50/50 chance of being, some time early next century, either side of 5°C above 1850 as a benchmark date. These aren't my probabilities, they come from the modellers. Five degrees is absolutely enormous. It would redraw the physical geography of the world.

Large parts of the world would become desert, including most of southern Europe and the southern part of France. Other areas would be inundated. You'd see massive movements of population. If we've learnt anything from the last 200 or 300 years, it is that big movements of population have a high probability of conflict. This isn't a black swan, a small probability of a big problem; this is a big probability of a huge problem. It's very important to be clear on that, because the DICE model has something like a 5 per cent output reduction for the world for a 5°C increase. That seems to me to be barely credible as a summary statistic for the kind of risk that I've just described.

The second point is how much you care about the future, and I think the authors of the DICE model get the discounting story wrong by attempting to infer discount rates from market evidence. I'm deeply sceptical about whether there's any credible outcome from that kind of argument. Let me tell you why. First, we must be very clear that discounting – how you value the future relative to now – is clearly of the essence. Focus on a single discount rate is obviously far too crude for this kind of story, as the scale of potential impacts means that discounting is very powerfully endogenous. We are not fiddling at the margins – the future path for the world depends enormously on what we do, so you can't read it off from something; it has to be directly endogenous.

Second, if you try to read things off from markets, you don't find a market for the answer to the question: what should we do collectively for problems which cover 50, 100, 150, 200 years? Because those are the consequences of our actions. The relevant markets for answering the question that you're trying to look at – which is a collective responsibility over a very long period of time –simply do not exist.

Third, looking at very long-run data, if you take riskless real rates of return you get 1.5 per cent or so, and if you look at equities you might get 6 per cent or so, depending on the period and the country you're looking at.

Which of those would be relevant for discounting? It would be the former – the riskless real rate of return – because you're going to take expectations and deal with uncertainty *after* that.

And fourth, a fundamental criticism of that route is that this is not a one-good problem. We mentioned this in the Stern Review, but Thomas Sterner[2] has demonstrated the point very clearly and in more detail. As people become richer and environmental goods become scarcer, it seems likely that, rather than fall, their value would rise very rapidly. Thus discount rates for the two goods will differ sharply. The flow-stock nature of greenhouse gas accumulation, combined with the powerful impact of climate change on ecosystems and other environmental goods, will render many consequences irreversible. This means that investing elsewhere and using the resources to compensate for any later environmental damage may be very cost ineffective. For all these reasons the attempt to infer high, and largely exogenous, discount rates from market data is profoundly misguided.

I think that the right way to look at this problem is as one of risk management. We should describe the consequences of different paths as best we can, ask what the insurance policy costs, and try to come to a considered, discussed, reasoned judgement: 'Given those types of risks, given these costs of the insurance policy, this makes sense as a risk management policy.' Looking at all these risks I've just described – movements of people, world conflict, the cost of world wars in human history, the kind of world war this could be – I find that the extra step that estimates this cost to be equivalent to 7 per cent of GDP, or 6, or 3, or 10 per cent, doesn't cut the mustard. It moves you away from thinking about the magnitude of these consequences. So I prefer a risk-management story, an insurance story: saying that if you can lower the probabilities by this amount for this per cent of GDP, with different percentages and different amounts of GDP, then we can find a good deal from the point of view of risk.

The two issues – (i) how big are the damages?; (ii) how much do you care about the future? – are dead centre. We have to discuss those directly, and we have to look at the ethics. On that, we have no monopoly as economists, but we have something to contribute.

Martin Weitzman (Harvard University) If you ask me why this world view has prevailed among economists – and this is *still* a majority view, although it's being eroded – I think it harks back to the way this was first conceptualized and the way it developed. It started in a deterministic worldview. It was phrased as an optimal control problem with a deterministic mechanical structure, which has lots of lags and complications but will translate flows into stocks, into heating the upper ocean and the lower ocean, and so on. And it comes up with a deterministic answer. Uncertainty was dealt with, in bits and pieces, by starting with parametric analysis to make sure things looked robust, and then kind of grudgingly doing something like Monte-Carlo simulation. But it never quite shed this philosophy that at the core it's deterministic. It never jumped into the uncertainty part with any enthusiasm, with a fresh view of what that might do. So I share with Nick, this is a tremendous experiment. Looked at geologically, we're way outside the range of experience. So I think the origins of this view were in a deterministic view of the world.

Thomas Schelling (University of Maryland)   Let me explain what Bill Nordhaus's DICE model is. It's essentially a model of an immortal entity that is deciding how much to invest to prevent some kind of potentially serious climate change in the future. And he uses a discount rate. (I think a lot of the criticism of your work [Nicholas Stern] has to do with your use of a low discount rate.) Bill Nordhaus asks, 'If I worried that 100 years from now my income would be drastically lowered as a result of climate change, which is the better investment: to invest a trillion dollars every 20 years to reduce the damage of climate change, or to invest a trillion dollars every 20 years in order to increase my income in the future?'. And he comes out essentially saying that at a market rate of interest, maybe you're better off investing in future income rather than investing in reducing the damage that you might otherwise suffer. Now this is not unique to Bill Nordhaus. There are a large number of economists who essentially look at this as a personal question of investment: which is better, to invest to increase my future income or to invest to reduce the future damage? Using an interest rate like a market rate, they essentially come up with the answer that you're probably better off investing your money in the future rather than reducing the potential damage. That result derives essentially from the use of a market rate of interest.

There are other people who use a discount rate with a pure rate of time preference, that is, almost everybody prefers earlier rather than later consumption. I find that irrelevant here, because hardly anybody displays a preference for earlier rather than later consumption by *somebody else*. In fact, most parents try to protect against the future profligacy of their children who may have this pure rate of time preference by arranging for their children not to get a lump sum, but to get a graduated income. And it used to be, at least in America, when families were more paternal oriented, that the life insurance of a husband was arranged so that in the case of his death, his wife wouldn't get a lump sum which she might squander, but she received a graduated income. So I think that it's clear that this pure rate of time preference is exclusively related to a person's own consumption. I don't think that there's any evidence that people prefer a benefit or a loss to occur 100 years from now to somebody they don't know rather than 150 years from now, so that I think that the pure rate of time preference has no relevance in this context.

The other thing is that as incomes grow, the marginal utility of consumption goes down, because the wealthier you are, the less benefit you get from a particular increment in consumption. Therefore, according to Nordhaus and so many other people, since incomes are likely to rise everywhere, the marginal utility of consumption will be going down. What they neglect, I believe, is that almost anything that is done to reduce the damage of climate change will be financed by the rich, and most of the benefits from reducing climate change will accrue to the poor. Therefore, the transfer is not from today's people to the richer people of tomorrow, but from today's rich people to the poorer people of tomorrow, who, though they will be less poor in the future than they are now, will probably still be less rich than the rich are now.

So if you use this notion that the appropriate discount has to do with the comparison of today's and tomorrow's marginal utilities of consumption, you have to recognize that we're talking about today's marginal utility of consumption of the people who have to pay for this and tomorrow's marginal utility of consumption of the people who will primarily benefit. Those are going to be the people in India, China, Bangladesh, Nigeria, Brazil, Indonesia and so forth, who will still, in 75 years, not only be poorer than the people in the rich countries, but will probably not yet have quite reached today's level of income in the rich countries. Therefore, the notion that this is a kind of personal investment strategy – how would I spend my income if I thought I would live for the next few hundred years in order to maximize my welfare 50, 100 or 150 years from now – neglects the fact that this is essentially a transfer of wealth from the people who can afford it to the people who most need it. And therefore I find the Nordhaus approach not to be pertinent.

Solow I think that we are agreed to proceed on the alternative view that has been so well expressed in the Stern Review and the other contributions of this book.

The second issue that I thought the panel might discuss begins to centre on what to do, what kinds of actions to take. The issue of policy directed at dealing with the prospect of climate change is different from many policy issues, in that it really is dominated by the spread of a probability distribution. Leaving aside the questions of radical uncertainty – which are very important, but which can be added in later – *at best* you're looking at, associated with any given set of policies, a wide frequency distribution of average surface temperature 100 years from now, let's say. And so we are asking ourselves to make policy decisions in this kind of environment, rather than in cases where the range of possible outcomes, or outcomes with reasonable probabilities, is quite narrow.

I think part of the difficulty of formulating and getting policy action here comes from that. I didn't much like earlier discussions about not knowing, and not knowing that we don't know. After you spend an hour talking about not knowing that you don't know, it seems to me that you've demonstrated that you *do* know that you don't know!

Leaving the radical uncertainty aside, let's think just about a wide range of possible outcomes, from a 1°C increase in average temperature to a 4 or 5°C increase or more. I want to ask whether anybody in the panel can think of other examples where systematic policy has been made success-fully in cases where the range of outcomes with reasonable probability for any given set of policy actions is very wide. Do we have any examples like that?

Stern I think before making any analogy we have to be very clear that we're playing with the planet, with temperature changes of 5, 6 or 7°C. You are not just playing with southern England or New Orleans, where if you define any particular place in the world, it'll be a small percentage of the total, and so you can say, 'Well, we can do without that'. In this case these are consequences that cover the whole planet. I agree with Tom [Schelling] that it's the poorest that get hit earliest and hardest. But at these kinds of temperature increases, we would all be very badly affected some time in the first part of next century if not before. This would either be direct impacts, through the places where we live being submerged or desertified or subject to hurricanes, or indirectly through the consequences of mass migration in the world and conflict. I think the planetary nature of the kinds of threats that are involved at 5, 6, 7°C drive this story very powerfully, and mean that small or even big probabilities of losing the whole planet are what we should be focusing on.

But if you look at examples, you have earthquake legislation in parts of the world that are deemed to have some probability, albeit low for any decade in a particular locality, of experiencing an earthquake. Much of drugs policy is pretty risk averse. We had some spectacular cases, such as CJD – mad cow disease – in the UK. They had a rule which said that you couldn't sell the carcasses of cows that were over a certain age. I can't remember all the figures, but the cost–benefit of that issue would have implied a huge value on life. In this case such evidence would suggest that it was probably a bad policy as a great deal was invested just to reduce the probability from a tiny fraction to a still tinier fraction for a particularly small group of the population when many more lives could have been saved by investing elsewhere. So in health, in drugs, in earthquakes and so on, there is this kind of legislation which is primarily driven by contemplation of very bad outcomes. Some of it is good policy; some of it is bad. I think that climate change is a much more serious example because of its scale and because of the rather high probabilities of catastrophic outcomes.

Solow   That's exactly the sort of thing that I wanted the panel to comment on. Another issue is that part of the problem of making policy is simple unwillingness to believe in the possibility of the worst outcome. That does not strike me as strange, as unusual, at all. You probably know that in the early discussions of the St. Petersburg paradox, there were proposals to resolve it by asserting that everyone treats probabilities smaller than a certain number as zero, and if you simply do that – as we all do when we cross the street – the particular problem disappears. But Nick made the very good point that there are two ways in which bad policy can be made. One is by focusing too exclusively on the bad outcome and reducing a negligible probability to a doubly negligible probability, at some cost, and the other is by refusing to face the bad possibility, and that's the sort of question I was thinking about. Does anyone have any comment to make on that?

Inge Kaul (Hertie School of Governance, Berlin)   There may be some people who have a problem in believing, I don't doubt that. But on the whole I think that the hesitation that we see in coming to action is related to global inequity, economic and political.

In international negotiations countries sometimes try to get a desired policy outcome the cheap way; and one of the 'cheap' ways of dealing with climate change is the current donor practice of debiting some of the related costs to foreign aid. Now, put yourself into the shoes of an African or Asian or Latin-American state. You have been promised 0.7 per cent (of GDP) as aid since the 1950s. That percentage has not yet been achieved. But in the meantime, it was being promised for this or that purpose – for literacy, health, and so many other things. And suddenly we're saying, 'Or maybe we'll use it for adaptation to climate change'. This is not an attractive incentive for developing countries. Aid money is limited, and the amounts involved are usually too small for the deep reforms that are to be undertaken, be it for adaptation or mitigation purposes. Today, 30 per cent of foreign aid flows into global issues, that is, issues that benefit not only the poor, but also sometimes even mainly the rich.

So we should undertake not only global cost–benefit analyses of climate change, but also disaggregated ones, and then see who the winners of a future global deal would be, what scope for compensatory transfers to the potential losers, and then construct win–win bargains. Finding win–win solutions requires political will and the recognition on the part of the winners that under conditions of policy interdependence, national interests may often be best served through successful – fair – cooperative policy approaches. Given the seriousness of the climate change problem, today's richer, industrial countries are often too cautious, maybe even miserly, in what they offer as incentives and compensatory finance to developing countries. This may look efficient in the short term, but it will prove to be ineffective and inefficient in the medium and longer terms, because follow-up to cooperative agreements will most likely be weak – making everyone worse off.

In order to speed up corrective action despite all the political problems we still face, it could be useful to establish an inventory of more-or-less immediately implementable, no-regret measures that we might want to undertake in any case, and that are leading in the right direction towards mitigation of and adaptation to climate change. Enhanced building codes are an example. But the scientific evidence suggests that meeting the challenge of climate change requires more far-reaching and costly reforms – with countries being affected differently by the different aspects of this multifaceted issue. All countries will have to contribute, especially those that have contributed most to the problem. Therefore, climate change is not a foreign aid issue alone. It would be better to agree on emissions caps, design well-functioning carbon-trading schemes and then let the carbon markets work.

Thomas Sterner (University of Gothenburg)   There's another way of thinking about this. There's a considerable difference in the way you deal with something like a serious cancer if you are the manager of a large hospital and have to plan ahead for resources, or if that manager himself has a serious cancer and has to take the decision for himself. I think we all know intuitively that risk aversion is going to play a different role and that the calculation is going to look different. This time, we are at the world planning level, meeting a problem as if we were a patient. Normally planners deal with environmental or medical problems in a statistical sense, but this is not a problem that can be dealt with in that way.

Solow   Does that make policy easier or harder?

Sterner   It should make us more risk averse. This is a way of conceptualizing what the precautionary principle means. For example, we can play around with rivers, and if we destroy a few rivers, then that is sad for the people who live along them, but there are other rivers . . . There's only one atmosphere and we use it all the time. So, it is obvious that we should be more risk averse in this case.

Solow Somehow that seems to me to be mixing up the range of alternatives with attitudes towards risk. And maybe there's something to that. Apparently losing a river or losing New Orleans is, in the light of this discussion, not of very great importance, but at least the corps of engineers is supposed to be thinking in precisely those terms, in protecting against the flood level that is expected to occur once per 100 years.

Sterner In that case they have to weigh the costs and benefits, because there is a budget restriction for the corps of engineers, and I imagine that the corps of engineers has other tasks. When it comes to climate change, it isn't *one* out of very many risks.

Solow But there are felt budget constraints, as Inge was talking about just now. Why don't we then try to face the issue head-on? What do the people on the panel think are the prospects for an intelligent, unified response to the climate-change threat? And after thinking about that, or maybe as part of it, what are the first steps that one should aim for, the steps that combine both adequacy and feasibility?

Kaul As we have seen from this discussion, the debate on the economics of climate change between Yale and London is clearly not yet settled. The controversy about this issue becomes more complex when we consider not only the views of different economists but also the differing views of Northern and Southern policy makers. Even if Nick were to incorporate into his studies some of the suggestions made by the previous panel speakers, the result would still be a *global* social discount rate, and a *global* social cost–benefit analysis. Yet we live in a world of wide economic, social and political differences and disparities. Therefore, an important question is how the net benefits, the costs and the benefits of inaction or of taking corrective action are distributed across the world. In other words, it would be desirable to have global and disaggregated cost–benefit analyses.

As the ongoing climate change negotiations indicate, delegations' views vary widely about which type of action might be warranted. The reason is that different groups of countries or even individual countries come to negotiations with their own cost–benefit analysis in mind. They all agree something should be done. But the preferences to act and to contribute vary.

This raises the question of the right approach to international negotiations on climate change. When governments appear internationally, they appear as national, and hence, quasi-private actors, pursuing their particular national concerns. So what has to happen in international negotiations like those on climate change is that the different parties have to enter into a bargain, a trade.

What struck me during all those years that I spent in the UN system is that this political market – international negotiations – is plagued by all the problems that we would no longer allow in economic markets. There are oligopolies of power, there are information asymmetries, there is free-riding when it comes to global public goods, and so on. So in order to move forward in a pragmatic way, we should look at international negotiations as a *political market*, a place where national policy reforms or other national inputs to international initiatives are being exchanged.

And if the international political market were competitive, Thomas [Schelling], you would indeed see a considerable transfer of resources from the currently rich to the poorer countries in the climate-change area. So be it! Why not? If you take historic emission levels into account, the richer countries have to pay off a certain amount of environmental debt, and if the price is right, if cooperation is attractive, I'm pretty sure we would see progress on the actions that go in the right direction, which are not drastic, which are not waiting, but which are actions that, for the time being, seem to be attractive from an economic point of view (or from any other beneficial point of view) to a number of countries, and we could move forward on that.

At present, some 190 countries and countless non-state actors are involved in climate-change negotiations, and therefore I was wondering whether we could facilitate the bargaining process by shrinking the 'market' platform. The question I would like to leave with you for discussion later therefore is: why have we turned to the G20 for leadership in solving the current financial and economic crisis? Why not involve this group in figuring out what are the basic differences in preferences to act on in the climate-change area? Maybe if this G20 met at the leaders' level, they could find some kind of cross-negotiation: some trade deals, patent-right concessions or financial transfers in return for climate-related national policy reforms. Maybe Nick can take the idea home to insert climate change into the forthcoming G20 meeting in London.[3]

In brief, I suggest moving away from a purely technocratic approach to the issue of the economics of climate change – with only the economists discussing it – and open it up to a more participatory political arrangement, where there is give and take. Moreover, we need to complement global cost–benefit analyses with country and perhaps even actor group-specific analyses so that all parties know how they are affected by the problem and by various proposed corrective policy measures. With these changes in place, we may get closer to getting the prices right. We would perhaps see more incremental change that may also help us to solve some of the uncertainties or unknowns that we see at present.

Stern I've been thinking about a global response intensely for a couple of years, and I've been talking intensely with those who are negotiating and those who have the responsibility of drafting the treaty of Copenhagen, but I haven't been negotiating. I stopped serving as a British civil servant 18 months ago.

I think a global deal looks like this: at least 50 per cent reductions for the world as a whole by 2050 and at least 80 per cent reductions for rich countries by 2050, all relative to 1990 levels. I'll state it, and then I'll discuss the challenges of getting there. The next part is that everyone recognizes, including the developing world, that by 2050, 8 billion people will be living in currently developing countries, and 1 billion in currently rich countries, with a reasonably high probability of those numbers. That is essentially a story about a world that is dominated numerically by the developing world, even more so than now. Even if that 1 billion in the rich world reduced their emissions to zero – spreading the total amount of emissions over 8 or 9 billion – doesn't make all that much difference. I think global emissions trading will be very important as a key element of the financial flows and the glue that holds the global deal together, as well as giving some notion of quantity certainty; nothing's for certain here, but it focuses on the importance of quantity certainty in emissions flows. We have to stop deforestation, which is around a fifth of current global emissions, and that can only be done in a way that combines the preservation of the forests and the economic development of the peoples of the countries in which the trees stand. We have to advance quickly on technology and share technology, and we have to consider the adaptation costs which will be substantial and particularly costly in human terms for the developing world, even if we are pretty responsible, as a major issue of equity. So those are the six parts of a global deal.

As for the criteria, Bob, I would add to your criteria of adequacy and feasibility: it should be effective on an appropriate scale, it should be efficient, keeping costs down, and it should be equitable, with a number of dimensions of equity. So I think that's what a global deal looks like. Now, why these numbers? Fifty per cent cuts by 2050 (depending, of course, on the whole path, but let's just talk about 50 per cent by 2050) would correspond to preventing the concentrations of greenhouse gases in the atmosphere rising above 500 ppm CO_2-equivalent. They are around 435 now. We're adding about 2.5 per year, so in about seven years or so we will be over 450. If you hold it to 500 and then gradually come down, or come down more rapidly if you find big ways of taking it out – which is an issue – then you have a chance of really reducing the risks that I described. We'd be reducing a 50 per cent probability of going above 5°C – under some notion of business as usual going on until the end of this century – down to about 3 per cent. There would be a major reduction in that probability.

There are many people who think we should be much tougher than that. Economists often think that I'm rather radical on climate change, but most of the environmentalists think I'm a bit weak and economist-like on these issues by just talking about 500 ppm. That would mean that 1990 emissions of 40 gigatonnes CO_2-equivalent would come down to 20. There'd be 9 billion people in the world, so emissions would be just over 2 tonnes per capita – remembering that giga and billion are the same thing – so 20 gigatonnes divided by 9 billion people is close to 2 tonnes per capita. At present, the United States is over 20, Australia is over 20, Canada over 20, most of Europe 10 to 12, Japan something similar, China 5 to 6 tonnes per capita, India about 1.5 and most of sub-Saharan Africa less than that. So that's the current distribution. Now, it's perfectly feasible to get the rates down in those ways, and we're getting better and better at understanding the costs of how to do it, but we must get much better.

We should recognize that the kind of programme I've just described is fairly minimally equitable. Eighty per cent cuts by Europe takes our 10 or 12 tonnes per capita down to 2 tonnes per capita. That would be saying that Europe, if it achieved the 80 per cent reduction that it has promised between 1990 and 2050, would be getting down to the world average by 2050, having been above the world average for 200-years or more, if you date the flowering of the industrial revolution to 1850. So in 2050, after a 200 year-old party, the Europeans would decide that it has finally got to the stage that we'll be drinking out of the same-sized glass, with no recognition in that story of all the guzzling that it had done for the previous 200 years of the party. What I've described is seen by many in the rich world as too radical in terms of action. My own view is that this is about emissions *flows*; it's not about emissions *rights*, and emissions rights are a very different thing. And there's arguably a case that rich countries should have zero or even negative emissions rights for a while and we can discuss that.

I wanted to sketch out what the global deal looks like, because these are the kinds of discussions that are taking place. I spent last week in Poznan at the COP 14,[4] which is the 14th Conference of the Parties to the United Nations Convention on Climate Change. The per capita story I'm describing is very much at the heart of those discussions – it is a political reality. I actually think it's a political reality with an ethical foundation, but it's a political reality that is a key part of the global deal. Now what could get us there? Well, after a few thousand people meeting for two weeks at Poznan, they came to this fantastic decision – and you'll appreciate the 'Yes, Prime Minister' feature of this – wait for it . . . 'It is time to move into negotiating mode!'. Well, they did take that decision. But it actually matters, because it means that all the countries there, including the United States, agreed to accept a draft text as a basis for discussion in June next year [2009], six months before the Copenhagen conference. It was an important move forward.

This is all tremendously dependent on two countries: there's the G20, there's the couple of hundred signatures of the UN triple-C (Framework Convention on Climate Change) and there's the G2! G2 is the United States and China, and their approach is absolutely fundamental. In Poznan, lots of developing countries came through with their climate change action plans, and this is very significant. They said that these were things that they were going to do anyway, *not* treating this as a one-off game with prisoner's dilemmas and free-riders and so on. Another huge part of Poznan was the election of Barack Obama, who has declared for 2050, relative to 1990, cuts of 80 per cent, and getting back to 1990 levels by 2020. That's a big task, as the US was 17 per cent above 1990 levels in 2007. It would be good if that could be upped to 5 per cent reduction by the US relative to 1990 levels, which is about 19 per cent off where the US is now. It's a big task, but it is feasible. That could start to bring the US into a position where others would feel it's starting to take a leadership role, and comparable actions elsewhere would put the world on a credible path to reducing emissions by 50 per cent on 1990 levels by 2050.

As for China, President Hu Jintao in a speech in June 2008, spoke of the importance of low-carbon growth. China is moving into its twelfth five-year plan, which will be starting from January 2011. The Chinese finish their plans well ahead of the beginning of the plan, whereas in India the five-year plan is published about a year after the plan begins. The discussion on China's future and the future of low-carbon growth will be taking place next year [2009].

I think there is now a basis for Barack Obama and Hu Jintao to get together some time in the first six months of next year and compare notes. Where are you going? Where are we going? What could you do, if we did a bit? I think we're almost ready for that kind of discussion, where it would have been inconceivable a year ago. Will that discussion have a strong outcome? Will Barack Obama be able to take whatever kind of deal there is through the Senate and the House? That's something that a lot of us worry about, but we *are* in a different position, both in terms of the way in which the United States is seeing and saying, and also in terms of the way the developing countries are now coming up with their own climate-change action plans. That includes China last year [2007], India this year [2008], and Brazil published its plan in early December [2008]. So we've got a chance. The human race is pretty skilled at blowing decent chances, particularly when it comes to cooperation. I don't want to guess the probability, but I would say it's a good deal higher than it was on 3 November, and it's much higher than it was two years ago.

Solow   Martin, is that the way you would go?

Weitzman   I don't know how I would go. In terms of policy, I'll share a few thoughts, for what they're worth. I don't have as optimistic a view as Nick does. I naturally think mine is more realistic. I'm in the sad position where I can't offer something more constructive or more realistic, so at the end of the day it's going to end up with, 'Hope for the best, but prepare for the worst!'. I notice in all these things that, yes, the rhetoric has gone up: there's more awareness; there's more words spent; the word-count of major newspapers and major leaders show that. But I also note that the specific plans that are put in place seem to be quite far in the future, not vastly far in the future, but 20 or 30 years off, typically. And they move, over time, to 20 or 30 years away.

I'd like to go back to a point that Tom [Schelling] raised that I really think is relevant. Whether it's tradable permits or anything else, they're not effective unless they're de facto putting on a very high tax. They're not effective unless people are feeling that they are impoverished, to some degree, by this move, and there is no getting away from that. That's my opinion, that the public will not be fooled for long. So all these agreements are dancing around. There's an idea that somehow the companies will take it on themselves. In my opinion, a big reason why tradable quotas are more appealing from a policy position is that they hide the costs, in the first round, anyway. To be effective, they must sell at higher rates that are going to raise fuel costs and everything else, so this is what I am not seeing – an awareness of how much this is going to cost, an awareness of how none of these new technologies, solar technologies and so forth, are in anywhere *near* an off-the-shelf state. They seem to me very pie-in-the-sky. Now I know the standard arguments, that if only we research enough . . .

I was talking with a reporter from the *Harvard Crimson*, who was asking me my opinion on Harvard going green, which is the usual stuff, 'We'll change the fluorescent light bulbs, grow grass on the roofs, whatever . . .', all of which is going to do nothing, even if it was adopted on a world scale. I shared with him my mixed feelings about ideas or plans that promote wishful thinking. I'll take Al Gore as an example, although I could take many others.

There's a message there that we can do something – of course it's symbolic – that we all in our lives can do *something*. These Harvard students are expressing this. We can do something. So this part is positive, that as individuals we can make a statement, we can get something started, it has propaganda value, it may inspire people. The bad side of that, though, is that it propagates the belief that this is somehow easy, or that this is somehow cheap, that it is somehow a question of awareness, of discovering many win–win situations that are out there. And that, I do strongly feel, is detrimental, because it puts off the real issue here.

I was touched by this reporter saying, 'You know, I'd never thought of it that way'. He was somebody who was in favour of all these Harvard green things. He said, 'That's right, it is intoxicating to get into this mould, to think that, well if we just do enough research this stuff will come on-line, it will be relatively cheap'. So I have genuinely mixed feelings about selling things that look too easy.

On the good side – which probably predominates if I weigh things up and if I'm forced to choose – it probably is good because it gets people aware. But it does have this bad side, which is negative, I think, of getting people in a mould where they think this is somehow pretty easy. It's *not* easy, it's very expensive. That's the core of the issue or the problem, and my own opinion, and in this I'm in a minority, is that it is better to level with the public somehow, whoever it is: this is going to be expensive. And that's one of the reasons, although there are many others, why I would prefer a tax over tradable permits in this area. In my judgement, we are better off in this particular instance telling the public the truth and letting them decide, as it were, than we are in trying to hide this or modify it so that it's less apparent. These are my few meagre thoughts.

Solow Tom [Schelling], I'd be very interested in your view, of whether a grand bargain of the sort that Nick was talking about, or an attempt to find a bargain like that, is more likely to lead to where we want to go than is a piecemeal approach – doing one thing and then when that is done, doing the second thing, and then the third thing . . . or any thoughts that you have about how to get started on this issue.

Schelling I have one agreement with Nick, which is that things are more promising now than they were on 3 November, or two years ago. I think that with the election of Obama we're likely to be in a position to think constructively.

I don't think Obama getting together with Hu Jintao is going to get anywhere. Obama's got to make sure he's got both parties in both houses of Congress with him, and I think that to go and negotiate with the Chinese by himself or with his new Secretary of State would be a mistake. Until the Congress is ready to send either the committee chairman dealing with this issue or the leaders of the two parties in both Houses – until they're ready to go with him wherever he wants to go – he doesn't have anything to promise; he can't negotiate. He has no basis until he has the Congress with him. I'm not sure that China is the place to begin. I think that maybe in five years China could become convinced that the United States and the European Union are serious. I don't think that the President going off and saying, 'I'm serious', will persuade them.

You referred to a global trading scheme. You can't have a global trading scheme unless you have something to trade, and the only thing you can have to trade is some kind of enforceable quota system, and I don't believe it's ever going to be possible to have an enforceable quota system, so what's to trade? I think announcing a goal for 2050 – I think this is consistent with what Martin just said – I don't think anybody can take it seriously, because nobody will know what it means, nobody will know what it will cost, and nobody will know what kinds of sacrifices are behind this commitment.

If you were to promise what you would do in the next five years, not in terms of emissions but in terms of what you would actually *do*, then maybe somebody can look and see whether you are doing what you said you would do. But if we have a target like, 'Reduce by 80 per cent by 2050', who will know for the next 10, 15, or 20 years whether we are on target or not? And when I say who will know, I mean who in the US government will know, let alone who in foreign countries will know. So I'm afraid that announcing a radical target for the future won't be taken seriously. You've got to be able to say, this is what it's going to take, and I think, as Martin was saying, you've got to level with the people who are going to pay the price.

If the US President could have a plan to get there by 2050, I don't think he could have a plan in the first year or two. It will take more time than that to draw up a plan to get to 80 per cent reduction in 2050. If he has a plan, and then if he follows Martin's advice – namely levels with the public, lets people know what their President is committing himself to, or trying to commit himself to – then maybe we could get somewhere, then maybe we could persuade the Chinese that we're serious. I don't think they believe we're serious. I don't believe Obama can persuade them, six months from now, that we are serious, because 'we' are the Congress and the people, and not just the President of the United States. So I'm very sceptical that big announcements for 40 years from now will make any significant difference, and I don't think a global system has any great chance of being brought about. I mean, who cares about whether Albania or Guatemala or Namibia or Tanzania or Bulgaria is enforceably committed to something?

I think what we need is for the main producers of carbon dioxide to be able to think about whether we can incur commitments to what we will do, not to the results in terms of emissions, but to what we will *do*. That means legislation, that means taxes, subsidies, programmes, regulations. What are we going to do about, as somebody mentioned earlier, insulating window glass, for instance? These all require figuring out how to make things happen and who's going to pay the cost. So I think announcing a 2050 goal is a way to evade the issue, rather than to focus on it.

Solow So if you were Obama's climate change adviser, what would you advise him to do first, whether it's in six months or a year or more?

Schelling I'd say get a team together to think about what legislation is going to be needed in order to demonstrate a firm commitment to doing something about climate change. Then get together with the leaders in the Congress to talk about what in the world they may possibly be willing to commit themselves to. Then think of a propaganda campaign to persuade the American people that this is very important, and it's going to cost a lot but the benefits are going to make it worthwhile. And if that takes him a year or two, do it! Then go talk to the Japanese, the European Union, maybe the Russians, bring in the Canadians, the Australians, and talk about what we can commit ourselves to do and if we *can* commit ourselves to do something, and *then* we go to the Chinese and the Indians and the Brazilians and the Indonesians and we say: we hope we have convinced you that we are serious. If we have convinced you that we are serious, we'd like to talk to you about how we can help you financially to do what you need to do in your own self-interest.

Solow Inge, how would you want to begin the policy issue?

Kaul I think there is a place in the world for what Nick calls a 'global deal'. Many people would probably like to have such a vision, whether others take it seriously or not, but there is an expectation that such a vision is being formed. However, as indicated earlier, I would not call it a global *deal*, because what is the *quid pro quo*? Global goals are being set. But many policy makers, especially those from developing countries, will ask, 'At what cost to me, my country, and what do we get from you, the industrial-country parties, as a compensatory measure if we take corrective actions?'. For it to be a deal, it should be clear what's in it for each country and whether it's going to be a fair deal. Now, how could we figure that out?

Climate change is a bit like poverty reduction. For the last 50 years we have wanted to reduce poverty, but how can you grasp poverty? You can't. So we have to disaggregate the challenge. In the case of poverty, the sub-issues could be health and employment creation, and so on. In the same way, to deal with climate change, we must, for example, focus on the challenge of reducing emissions and discuss whether to go for emissions trading or taxes. But there is also the question of energy security, a challenge that requires quite different and separate policy choices. For example, we would have to look at patent rights, and what sort of changes we make in TRIPS [Trade-Related Aspects of Intellectual Property Rights Agreement] so that we can better disseminate available clean technology and enhance global static efficiency.

Therefore, my advice is to take the issue of climate change to the top leadership level – the G20 that at present is dealing with the global financial and economic crisis.

Moreover, under the main G20, which meets at the level of heads of state or government, other G20s – perhaps ministerial-level ones – could address various sub-aspects, such as a more flexible TRIPS interpretation to facilitate the dissemination of new technology in support of energy security. It is only when one breaks down big issues like climate change into their sub-components that the underlying incentive structures become apparent and possible attractive *quid pro quos* can be identified.

What has been so typical of negotiations in the past is that we agreed on ambitious objectives. Then we went home and forgot about it, and nothing much happened. Ten years later, we came back and formulated even bigger expectations. But challenges like climate change require that we move into a more executive mode, that problems get resolved. At present we only have a legislative part in international negotiations (apart from the limited foreign aid arm). Yet on many global issues we must achieve results, lest global 'bads' continue to adversely affect the world. Therefore, we need to involve world leaders and create issue-specific management groups in order to work out attractive global deals and move from words to deeds.

The incentive structures that need to be built, the compensatory financing that may have to be made available must be decided on an issue-by-issue basis, and so I would recommend looking at the possible policy package as a market deal – an exchange of policy reforms in one country against matching policy reforms or compensatory finance in others. The net benefit to everybody has to be clear. So my advice for dealing with the climate-change issue would be to follow an approach similar to the one the world has chosen in respect to the current financial and economic crisis: let's decide on an institutional arrangement that provides highest-level policy leadership and support for the deep and far-reaching reforms that climate-change mitigation and adaptation will no doubt require in all countries, rich and poor.

Solow What worries me about that, Inge, is that the answer to the question, 'What's in it for me?' is primarily that the sea level won't rise by 5 metres all along your coasts, not that we're going to give you a present, or give you something so that you'll give us something. The cost to the world is being expended in order to prevent temperature from rising and sea level from rising, and that's what the benefit is.

Kaul This is correct. But the political market, international negotiations, must be competitive, so that we actually get the 'prices' right, ensure that cooperation is the preferred policy path for everyone. Only then will we reach the desired outcome. So it's not only, 'that there must be something for me'. *All* parties must be able, in my view, to go home and show a net benefit from international cooperation. Therefore, the way in which we calculate the costs and benefits of inaction, as well as those of corrective action, is extremely important. The point I would like to emphasize is that there is not only one – global – way of determining the costs and benefits of climate change, not only one global discount rate. That is because people's and countries' conditions, and hence preferences, vary.

Solow But for most of the world, the net benefit of the thing, as I understand it, is that we avert climate change, or we avert so much of climate change – not that there is something extra to give around – it's that we ward off the problem.

Kaul Yes, the costs of global corrective action often are but a fraction of the costs of inaction, and taking corrective action may generate significant net benefits. But both costs and benefits are often unevenly distributed across countries; some nations may gain and some may lose. Therefore, in order to make international cooperation happen, I am suggesting finding ways of constructing a fair global deal.

Solow Thomas, what are your thoughts on this issue?

Sterner We are speaking of several big issues. I understand the doubts being expressed about targets for 2050 for the world as a whole, but then on the other hand it's a very exceptional issue, and that issue is tied to the carbon content of the atmosphere, so maybe there *is* a role for an umbrella-type of goal at that level. I think it's still important to bring things down and look at sectors as well, and it's illustrative to look at the one policy that actually has had an effect so far, which is gasoline taxes – one of my favourite subjects!

Gasoline taxes in Europe have actually reduced emissions, and I put some effort into calculating this. If you think in tonnes of carbon, you were saying that currently the world is emitting about 7 gigatonnes of carbon per year. About one of those comes from the OECD transport sector. I calculated, using the conventional elasticities, what this would have been if we hadn't had gasoline taxes in Europe and Japan. It would have been about 1.5 tonnes, so about half a gigatonne, which is almost 10 per cent of global emissions, has actually been cut just because of the gasoline taxes in Europe. A similar amount would be cut if the rest of the world had gasoline taxes. So gasoline taxes have quite a big effect – not enough to solve the problem on their own, but it is certainly a very significant effect.

Now let's look at feasibility. Gasoline taxes are considered unthinkable in the United States, but if we ask an audience here in Europe, which already suffers from gasoline taxes, how high up do they figure on the list of things that really burden them when they can't sleep at night, the answer is not very high, because we've adapted to them. And of course in reality, the fuel taxes replace other taxes and are recycled back into the economy. So in the end, living in a country with gasoline taxes is not so bad.

It might seem hard to raise fuel taxes in the US, but many policies appear before they are actually implemented. Who ever thought that Nelson Mandela would be taken out of prison and made President? That Barack Obama would become president of the United States? Who ever thought that in France you would put on seat belts and stop smoking? Nobody! This was unthinkable, just as unthinkable as the idea that the Berlin Wall would come down. There was nothing more solid than that Berlin Wall, and now it's just gone! So if I were to sound prophetic, I would say that one day the United States will have fuel taxes. This policy is simply effective and reasonable, and it is likely policy makers will come to realize this in time.

Now the trouble is that the shadow value of carbon implied in European gasoline taxes is too high, at least in the short run, to apply to heating and electricity production in India, for example, so unfortunately, I think we may have to temporarily relax the idea of having an equal price on carbon everywhere.

Solow   I wonder how Obama's advisers would feel about that. Remember, it's not as bad as it sounds initially, although in the US, of course, the political problem is that most of the gasoline taxes are in the hands of states, and for the federal government to pre-empt this would be a big change. But the proposal would be to increase gasoline taxes and reduce other taxes. You could imagine a revenue-neutral change.

Sterner   Yes, I think this would have to be done in all kinds of clever ways, sharing the burden and the gains between the states and the individual motorists and the federal level; there are lots of details there, obviously.

Solow   I wonder whether there's been any research in economics, using input–output tables, to try to estimate the changes in relative prices that would result from a carbon tax. Those are the main price changes. When Martin talks about bringing home to ordinary people that it's costly to have a carbon tax or for that matter a tradable permit system, what that really means is that the relative price of carbon-intensive goods and services will rise relative to the others.

Kaul   And that's the purpose.

Solow   Yes, that's the name of the game. But Nick, you wanted to reply . . .

Stern   I agree with a lot of what Martin and Tom [Schelling] said, but I also think it is very important to have an understanding of where the politics lies. I don't think it's quite in the same place, either in the United States or elsewhere as has just been described. Martin said we've got to level with people about the costs and we've got to be clear about the level of the costs – there's a double meaning of 'level'. What does it mean to level with people? Well, you do your best to calculate the cost, to calculate the kind of carbon prices that might be implied in these different kinds of policies; you look at the overall cost, it's not just the marginal, not just the carbon price, it's the integral along the cost curve that gives you the cost. And what do you get when you look at that?

Let me just give you some mental arithmetic. For the path I've just described, which as I said is much less radical than many environmentalists would like, but more radical than some economists, for a cut of 50 per cent by 2050 relative to business as usual – and there are the usual problems with defining business as usual, but put that to one side – you'd have to take out about 50 gigatonnes per annum. That would probably – and we need more work on this – but it would probably cost in the region of an average cost of about 40 dollars per tonne of CO_2. A substantial amount of that would be negative, a substantial amount close to zero, and a substantial amount well above 40 dollars a tonne. So 40 dollars a tonne times 50 gigatonnes is 2 thousand billion dollars – 2 trillion dollars – and world GDP in 2050 might be twice what we are now if we're sensible. If it's more than twice, splendid, but for the sake of this argument, doubling world GDP would take us from around 50 trillion dollars to 100 trillion, and 2 trillion would represent 2 per cent of that.

That's a back-of-the-envelope version of much more detailed cost calculations that you could make, based in terms of integrals along cost curves to get the average cost. My guess is that 2 per cent of GDP is an upper estimate. The IEA [International Energy Agency] has come in with lower numbers than that, McKinsey has come in with much lower numbers than that, and the IPCC [Intergovernmental Panel on Climate Change] with considerably lower numbers. I like to make an allowance for bad policy which can push the cost up. I've seen bad policy in action, so I always make an allowance for it.

Many people are trying to level on what those costs would be. It might well involve, in the near term, a carbon tax or a trading price, whichever, of around 50 dollars a tonne of CO_2, and that would be equivalent to roughly 25 dollars on a barrel of oil. These are significant costs, and we can and should and many of us, I hope, *do* level with people about how big they are. And then you can go deeper into them, look at the relative prices. The price of steel or aluminium could be going up by 20 or 30 per cent, whereas the price of many services would not go up by anything near 2 per cent. That's the kind of changes in relative prices that you're talking about.

They are significant, but not huge, not relative price changes of 100 per cent or anything like that, at least at broad levels of aggregation. So that's about levelling with people and estimating the level of costs. I don't think it's correct to try to fool people, and I react against the suggestion that that is what you try to do. Having said that, I do think that if you ask someone on the Clapham omnibus in London if there is a carbon tax, they would say no, when in fact we've got quite heavy taxation of petrol, and the European Union emissions trading scheme, currently at about 20 euros a tonne of CO_2, maybe 25 or 30 dollars a tonne of CO_2, covers well over 40 per cent of EU emissions. So I take Martin's point that they know more about the tax on petrol than they do about the European Union emissions trading scheme. There is a question of openness and I would, with Martin, go down that path. I think that involves explaining to people more about the trading scheme than we have and explaining how the price is passed through.

Schelling   Does anybody think that 25 dollars a barrel would really make a lot of difference?

Stern   I think you have to look at the way these trading schemes work. If you think of the big players – electricity, aluminium, steel – and you give people a carbon price, they will simply look at it as a cost-minimizing problem, 'Do I keep my costs down by going on these alternative measures?'. There is a whole swathe of promising reductions that are viable at 25 dollars per tonne of CO_2, which is where we are now, and many more that would be viable at 50 or 25 dollars per barrel of oil. This is not simply consumers thinking about the engine size of their cars. I think those kinds of decisions are much slower, although as Thomas Sterner was saying, the carbon-per-mile in Europe is about half that of the United States, and you have to believe that much of that is related to the taxation.

For the big bulk of the emissions – which takes place in big units where cost minimization is a perfectly reasonable assumption – a price of carbon dioxide of around 50 dollars a tonne would make a difference. But you're going to need regulations in some areas where the markets don't work so well. You're going to need to stop deforestation. You're going to need to think about how housing markets work and other failures beyond simply this one. I don't think that a price of 50 dollars now on the margin, 100 dollars perhaps in 2050 on the margin, is going to do it by itself. You need another set of measures as well, but I think it would make a big difference.

The second set of issues concerns timing, and I very deliberately gave the example of US targets for 2020, because still more important than the discussions between Poznan and Copenhagen during these next 12 months will be the 2020 targets. You saw what happened in Europe when the real bite of the 2020 targets started to hit people: we saw a difficult discussion last week [December 2008] where the targets were challenged, but, which nevertheless culminated in an agreement between the heads of states in Brussels to stick with the 20 per cent reductions by 2020 relative to 1990; that's the next 10 years, so we can see what we're doing there.

The UK has a target of 80 per cent reductions by 2050, and the climate change committee which holds the government to account on this has just published the detailed steps and the five-year budgets that are implied in a move between where we are now and those 2050 targets.

I was concrete about what a US target for 2020 might look like – only 10 years away – and Barack Obama has been quite explicit on where he thinks it should be. He said we should get back to 1990 levels by 2020, and that means a 15 per cent cut between now and 2020. This is talking about policy over the next few years, and that's where the discussion has to be. So I agree with you that that's where the discussion has to be, but I disagree with you about where the discussion currently *is*. Target for 2020 is where it's moving now and that is when, as you say in the States, the 'rubber hits the road'.

Now, as to where the US is in politics, you're American and I'm not, but I'm quite close to what's going on. I've testified twice on Capitol Hill, and I've stayed pretty close over these last few months. The Warner–Liebermann bill failed this last year, but it got a long way, and there was a lot of support. You now have a changed Congress and House; you have a President who is pro rather than anti; you've got Steven Chu in energy; you've got California; you've got a whole swathe of American cities that have made strong commitments on this. It's absolutely clear that Obama going to China by himself would leave himself exposed to all the kinds of risks that you mentioned, but what I disagree with is that that is a description of what is being suggested. I think we're seeing very intense political discussion by those who have to cast their vote in the House and the Senate taking place right now, and I think it will intensify.

Another question is how it stands in relation to the economic crisis, which maybe we can come back to. So I agree with you that you have to have more than just the President of the United States, but it *is* more than that and it *can* be more than that: it has to be built.

I agree with Tom [Schelling], that showing the credibility of the United States is fundamental. I believe that that will be built into any agreement in Copenhagen. In other words, countries like China and India and so on will commit, subject to the targets being taken on, subject to there being visible progress, subject to there being some kind of financing and subject to sharing of technology. I think that the best hope for an agreement is something like a conditionality put by the developing world on the developed world, saying that we like what you're saying. If this develops along these kinds of lines, then we're ready to make stronger commitments ourselves, but in the meantime here are the commitments that we are ready to make.

India, China, Brazil, Mexico – even more intensely – and many other countries are thinking it through. In those four cases I've been directly involved in some of their studies. They're trying to work it out: what does it mean for us; what actions do we take; if we all move together as a world, what kind of reductions in probabilities do we get? Those are the kinds of discussions taking place.

China in particular is concentrating very hard for two very good reasons. One, it's increasingly working out just how vulnerable it is. China is a nation with a big density of engineers at senior policy-making levels. They're looking at the water challenges, which are huge. Their main rivers, particularly the Yellow River and the Yangtze, rise in the same part of the Himalayas as the Indus and the Ganges, and so on. They're working it out: what it means for them, what risks they run, and they are very clear that they are very vulnerable. It is also very clear for China that they are a deal-breaker. If you are very vulnerable and you are a deal-breaker, you concentrate, and that's happening.

The international dialogue that we're discussing is moving forward. That's what makes me more optimistic. I don't want to be cast as optimistic; as is common amongst economists, I talk about rates of change rather than levels – but I would be optimistic.

Whether we get there, I really don't know, but I *do* believe that acting as if we've got a chance of getting there gives you the chance of getting there. If you act as if you've got no chance of getting there, you will not get there. At the same time, I agree with Martin, you've got to think about what happens if we don't get there, and that's rather worrying, but should not be a subject which is taboo.

Solow I'm a little worried, Nick . . . Obama did run for office promising whatever he promised in terms of carbon reduction, but he did *not* run for office by offering to give up say 3 per cent of GDP per year for the good of the world; he just ran for office offering to achieve the good of the world. And I think it will be much harder to get the kind of agreement that would be necessary to do the actual levying of taxes. Just harder.

Stern That's right. I've worked as a senior civil servant for politicians who make promises where you wonder whether they've been costed – actually, you know they haven't been. At the same time, this is a discussion that surely will take place, and must take place. I think Martin and I agree very strongly on that. I also think that those numbers are probably fairly generous estimates of the cost. The rate of fall of the cost of renewable technologies is really quite remarkable. Suntech – a Chinese company that is one of the biggest PV [photovoltaic] companies in the world – will be investing in southern Italy next year, with solar PV that's competitive with current grid prices at the point of delivery to the consumer (the costs of generation per se would be higher for PV, but it is the price at point of delivery that matters). These are the kinds of changes that are taking place. Now on average, solar is still expensive, but in some areas it is now cost competitive.

In the United States last year, in 2007, 35 per cent of the installed capacity was wind (the *new* capacity as a fraction of total capacity is tiny). These are the kinds of calculations that are taking place. With the kinds of incentive structures we've got at the moment, it's changing very rapidly. France changed from very little nuclear to 70, 75 per cent nuclear in about 20 years for electricity.

These kinds of things can happen very quickly. There are costs of doing them, but they're not growth-breaking costs. They are not costs that we are unfamiliar with from very big changes in exchange rates, and so on. We must face the discussion of what those costs are directly, but we have to do it in a careful, analytical and balanced way. The numbers I suggested are probably on the high side. But governments could foul it up sufficiently that they could be on the low side.

Schelling   Fifty dollars a tonne is about 15 cents a gallon. I don't think that makes a difference. If you get it up to 250 dollars a tonne, *that* may begin to make a difference, but 50 dollars a tonne is only 15 cents a gallon! And the same thing applies to electricity; I think it will be negligible in the United States.

Stern   No, the cost-minimizing entity for a big part of this – the people who are actually making these generation decisions – are things like electricity companies. If they see a price of carbon, and they see an alternative – a way of cutting back on carbon which can beat that price of carbon – why would they not do it? And those are the kinds of choices that are taking place right now in the European Union emissions trading scheme, which is still in its early days. You have to distinguish the kind of perception and calculation of options that take place in the cost-minimizing story with people who know about the technologies and know about the alternatives, and the consumer story, which would have much less information and understanding associated with it.

Weitzman   I have three points. I agree with what Nick said, that if you want to achieve good aims, you've got to try to do it, otherwise it's not going to be achieved. His role is very important, and he is the ideal spokesman: he is articulate, knowledgeable, forceful, so you are doing a good thing. My guess about what this will come to is another issue, but it's a very important role and you're ideal for it, and if you didn't exist we'd have to invent you. There needs to be someone doing this, simply for the reasons that you said, you're not going to get anywhere unless you try. You want eggs in that basket, maybe even significant eggs; what the probability is that they'll make an omelette is another matter.

The second thing is this: maybe I'm battle-weary, especially from my days at MIT where I interacted plenty with engineers, and I have been hearing this all my life, 'costs are coming down if only we wait'. It rings in my brain, and I could produce at least a dozen examples. Engineers are prone to that. I don't, reading this literature, see anything significant happening in solar; I don't see these costs coming down; I don't see it moving towards commercialization. It's still very far from that, so I don't perhaps put as much weight on these stories as you do. Engineers who tend to make these claims often get mixed up between what is technologically feasible and what, when you look at the details, can actually turn a profit. So I'm somewhat more sceptical on that.

Finally, my third point: this thing about the carbon tax and it being 2 per cent. Okay, I think along with Tom [Schelling] that it is in fact considerably higher when you do the calculation, but let's leave that aside . . . Suppose it is 2 per cent. Is that a lot or a little? Well, again I've been subject to many scientists saying, 'Hey, it's only 2 per cent, and I would very willingly do that; it's only a small change in the defence budget, or in the way we do this or the way we do that'. Yes, that's true, but I'm not sure that the world population shares my view on that. Two per cent is also a huge amount of national income (a trillion dollars). Here's an example, a place to attach this to, and there's many. If you take the world petroleum industry – all the oil that's extracted in the world – price it out at its current price and subtract from it its costs, *that* amounts to about 2 per cent of world GDP. So we think of this as a massively expensive big business operation: it is 2 per cent of GDP. So it depends on who is looking at it: 2 per cent of GDP could be considered a massive amount of money to ask people to forgo, or it could be looked at another way and seen as some minor reduction in military expenditure. All I'm saying is that there's this other side to the story that says that this is a lot of dough.

Solow   One year's growth is a way of making it look smaller.

Weitzman   Yes, that's right, and then it continues at 2 per cent, compounded.

Kaul   Whatever approach Obama might want to choose, he should not say that he's doing it for the good of the world *only*, because addressing climate change also serves the national interest of the United States. If politicians were to highlight more that international cooperation, if done right, can generate important national net benefits, it would probably be easier for them to win the support of their constituencies for these measures. And then should we call the financing of climate-change initiatives a cost or an investment? I think it would be much better to get used to thinking in terms of investment – spending money, including public revenue for climate-related measures, where it will generate a relatively good return for the country.

Solow   Investing in the future, that sounds good. Let's try that!

Sterner   That might be a way to sell it. But I thought it almost sounded as if my eminent friends were mixing up costs and taxes. To give an efficient signal we need a very high gasoline tax, but that money is not lost, it is just a signal, and it can be refunded elsewhere. It is just a device to change consumer behaviour, so we shouldn't mix those two things up. We Europeans are still here. On average we are not moving to the US just because gasoline is cheaper over there! It doesn't make that big a difference to our daily lives. A several hundred per cent tax such as we have on gasoline is not such a big cost in the end.

What is a cost, ultimately, is if we have to build wind-power stations, and they are more expensive than coal-fired stations. Now *that* is a real resource cost. It is that kind of thing that I think will amount to 2 per cent of GDP ultimately. But then if we're growing at 3 per cent – which we're not this year, but if we do resume that – then in a 50-year period we go from an income of 100 in the world to 400, and if we lose 2 per cent that means we go to 392 instead. We know from psychology that a gain that is smaller than it might have been is very different from an out-of-pocket cost, so I think that we're doing ourselves a disservice if we speak too much about this as a cost. It is true, as you [Weitzman] were saying, that we have to be honest with the public, but it may be more than honest if we say this is a 2 per cent cost. That is exaggerating the cost, because the public will think, 'I go from 100 today to 98 tomorrow'. They think, 'I've got a lot of fixed costs, so I lose all my pocket money'. That's what they think if you say a 2 per cent cost, but that's not what we believe. We believe that we're going to get richer in the future, even if we adapt to climate change. We're just going to get richer a little bit slower. That's very different. I think, it may not be all that easy, but if we manage to explain that, then it will be easier to get acceptance.

Stern   There are other benefits that come through much more quickly. We're all focused on the long-term benefit of reduced risk from climate change, but a lot of countries, and certainly Brazil in switching to ethanol and France in switching to nuclear, took the energy security issue very seriously. They thought that in large measure, they were not simply controlling their costs in anticipation of expensive hydrocarbons, they were also thinking through the challenge of energy security. This is a big issue of discussion in the US and Europe. So a lot of renewable energies, and nuclear, give you increased energy security, relative to the kinds of structures that we have at the moment. You also get less local air pollution, and you have a quieter structure generally. As part of all this we must stop deforestation – otherwise we'll never get there – and that will also bring greater biodiversity. So there are some benefits which most of us note, but which few of us quantify, which come through more quickly.

It's part of the story that low-carbon growth is the only growth that can sustain itself, because not only will high-carbon growth kill itself off early on in terms of oil and gas prices, but also before very long in terms of very hostile climate. So high-carbon growth isn't really a medium-term growth option. Low-carbon growth is. You pay a fair bit over probably two or three, maybe four decades (after that it may go down if we get clever) – the kinds of figures we've been discussing – but you also get some of these other benefits coming through in a comparable time frame, and most of us note them but few of us calculate them.

Solow   You would have no trouble convincing me. I was sold on the value of the nuclear option by very intelligent nuclear engineers who explained what they could do that was safe and intrinsically safe and so on. I have had a better experience with engineers at MIT than Martin did! Nevertheless, I visualized, as you were speaking Nick, a thought experiment. I collect all the mayors of communities in the US into a large room, and I announce that as part of the low-carbon growth problem we are at last going to build a nuclear power plant. Which of you would like it in your town? And the silence is deafening. Or each of them points to the next one in line. It will have to be done, I am entirely with you on that, but it will not be as easy as it sounds.

Stern   On that, we are in France, and many people know the details much better than I do, but there are deals, financial incentives for local communities to accept power plants.

Solow   Yes, but it is also true that throughout history the fraction of downtime in French nuclear power plants has been much smaller than the fraction of downtime in American nuclear plants. That may not be true any longer, but then they haven't built any in the last 50 years.

Weitzman   On this issue of the 2 per cent cost, I think maybe I was misleading: having heard some comments and criticism I want to change what the implication was. First of all I still believe – I'll have to look this up again – that it is more like 3 or 4 per cent. We don't know, of course, but it's higher than 2. Be that as it may, leave it as a minor correction, if there is such a correction to be made. I agree that that is not very much to pay; that is the proper way to look at it.

But somehow again there is this issue that I don't think we're going to get anywhere until the people start being levelled with. This is part of it, for the President or somebody to say, 'Look, this is going to cost in these areas; this is going to change lives in these areas; but look at it this way, it's only 3 per cent', or whatever. 'And look at it this way: we're trying to offset that by some tax adjustment on payroll taxes', or something like that. So that point *is* important, that it's not such a high fraction of GDP.

But it's necessary, I think, to sell this honestly. That's been sadly lacking everywhere. There's still this air of unreality, I believe, about the whole thing, in announcing targets far in the future that are then going to be revised. There's an air of unreality that needs to be removed in my opinion and replaced by this more honest approach, and then the case has to be made that way.

NOTES

1. The Dynamic Integrated model of Climate and the Economy (DICE) is the archetype of this approach, which has been elaborated by William Nordhaus and his team at Yale University. The latest version of DICE is published in *A Question of Balance* (Yale University Press, New Haven, CT, 2008).
2. Sterner, T. and U.M. Persson (2008), 'An even Sterner review: introducing relative prices into the discounting dabate', *Review of Environmental Economics and Policy*, **2** (1), pp. 61–76.
3. The G20 meeting took place on 2 April 2009 in London.
4. December 2008.

Index